CHEMISTRY AND POLLUTION

CHEMISTRY AND POLLUTION

edited by

F. R. BENN and C. A. McAULIFFE

*Department of Chemistry, University of Manchester
Institute of Science and Technology*

M

First published 1975 by

THE MACMILLAN PRESS LTD.
London and Basingstoke.
Associated companies in New York Dublin Melbourne
Johannesburg and Madras

SBN 333 13888 0

Filmset and printed by Thomson Litho Ltd., East Kilbride, Scotland

CONTRIBUTORS

Professor G. Scott — Department of Chemistry, University of Aston, Gosta Green, Birmingham B4 7ET.

Dr. A. Porteous — Faculty of Technology, The Open University, Walton Hall, Walton, Bletchley, Bucks.

F. R. Benn, Esq. — Department of Chemistry, University of Manchester Institute of Science and Technology, Sackville Street, Manchester M60 1QD.

Dr. C. A. McAuliffe — Department of Chemistry, University of Manchester Institute of Science and Technology, Sackville Street, Manchester M60 1QD.

Dr. R. Perry — Department of Civil Engineering, Imperial College, Prince Consort Road, London S.W.7.

Dr. D. H. Slater — Department of Chemical Engineering, Imperial College, Prince Consort Road, London S.W.7.

G. E. Eden, Esq. — Water Research Centre, Stevenage Laboratory, Elder Way, Stevenage, Herts.

Dr. G. Nickless — School of Chemistry, University of Bristol, Cantock's Close, Bristol BS8 1TS.

CONTRIBUTORS

PREFACE

Few people who live in industrialised societies can be unaware of the present concern about the predicament of the environment. Much has been written about the fate which awaits mankind should the advice of the more pessimistic ecologists be ignored; in fact, some say that the end of our civilisation is so close that little can be done to prevent catastrophe. The more optimistic commentators have emphasised that man's awakening to the dangers and the adaptability of technology will enable civilisation to continue.

It has seemed to us that, while a bewilderingly large number of books and articles have been written on the 'ecological crisis', these have quite often been of a somewhat superficial kind, and it is the intention of this book to present a factual chemical background to one facet of the ecology discussion—pollution.

Most pollution is chemical in nature, but it is quite obvious that although phrases such as photochemical smog, biodegradable polymers, hazards of DDT, are frequently mentioned, there are very few people who actually know how photochemical smog is generated and what is its composition, or for that matter, the chemical structure of DDT. There is thus a considerable gap at the very elementary level in the present wide discussion of pollution.

In order to fill this gap we present this volume, written by chemists who have backgrounds of teaching and/or research experience in the fields they have written about. The brief we gave the contributors was to describe the chemical (and industrial, where appropriate) basis for pollution of a certain kind. We asked them to restrict their discussion, as far as this was at all possible, to GCE 'A' level (about Freshman Chemistry level in U.S. universities). We made no attempt to restrict the authors' individual styles, and as most of the authors are actively engaged in research in the subject areas of their chapters, we feel that they have frequently been able to inject some of their own enthusiasm, and their chapters reflect the thrill of attempting to solve huge problems by chemical and technological ingenuity. Thus, although there is a small amount of non-uniformity throughout the book, we feel that this has ultimately made for a more interesting and informative volume.

The chapter entitled Polymers and Pollution points to the growing use of plastics in packaging, and to the resulting environmental problems of long-lived polyethylene and PVC. A comparison of the structures of cellulose and synthetic polymers gives some insight into the problems of disposal. Recycling, particularly topical in times of high petroleum prices and a growing awareness of the need to conserve materials, is discussed as well as the degradation of plastic wastes by bacterial and photochemical methods.

Domestic refuse, a problem which grows at an enormous rate both because of population growth and the growth of the individual consumer's wastes, is also discussed in terms of the tremendous recovery potential. Schemes are outlined for the production of oil and gas from domestic refuse, and the hydrolysis of cellulose ultimately yielding ethanol shows that the application of technology to pollution problems can not only alleviate environmental pollution but also contribute to the conservation of resources. The more conventional methods of dealing with domestic refuse, pulverisation, composting, and incineration are also detailed. The section on pesticides outlines the chemistry involved in the synthesis of the most common pesticides, DDT, γ-benzene hexachloride, aldrin, dieldrin, tetraethylpyrophosphate, and parathion, as well as the conversion of DDT to DDE. The inertness of the organochlorine pesticides is discussed, and an attempt is made to evaluate critically the hazards associated with these materials against a background of immense good which has accumulated to man through their use, especially in the underdeveloped countries. The chemistry of the very reactive chlorophenoxy acid herbicides is outlined, as is that of the triazines and paraquat. The chapter on air pollution reviews the types of pollutants present in air, the monitoring of these, and the known and possible effects of these on biological systems. A brief outline of the chemistry of short-lived excited species, free radicals, in the atmosphere emphasises their importance in the generation of certain types of air pollution. Pollution from exhaust systems of internal combustion engines is described, as are possible ways of avoiding these emissions. Sulphur dioxide, the United Kingdom's greatest air pollutant is mentioned in some detail. The composition of domestic sewage introduces the chapter on waste waters, and the currently available treatments, including water reclamation, are outlined. The impact of detergents on sewage treatment, and other problem cases, free and complexed cyanide, acids, and radioactive wastes, are discussed in detail. Detergents are also the subject of a separate chapter where the complex chemical constituents which make up household detergents are listed. The various types of surface active agents: anionic, cationic, non-ionic, and ampholytic are described, and the problems of eutrophication resulting from detergents is discussed.

We believe that the types of pollution dealt with give an almost complete coverage of pollution which affects us most. Pesticides, polymers, and photochemical byproducts are perhaps obvious subjects for inclusion; we have two chapters which discuss aqueous pollution—waste waters generally, and the more specific problem of detergent pollution; the chemistry of domestic refuse has been included, and we feel that this particular facet of environmental pollution receives much too little discussion.

We are aware that a number of universities and college now offer courses which include the subject of environmental pollution. We hope that by keeping the basic level of chemical theory elementary this book may be of use to students with only basic chemical backgrounds.

The individual authors wish to acknowledge with gratitude the patient way in which their wives forgo a good deal of companionship so that work of the type described here can be thought about, acted upon, and finally written down. We are especially grateful to our wives, Jill Benn and Margaret McAuliffe, for the help they have given us in this way.

F. R. Benn
C. A. McAuliffe

Manchester, 1974

CONTENTS

Preface

CHAPTER 1

POLYMERS AND POLLUTION

GERALD SCOTT

Department of Chemistry, The University of Aston in Birmingham,
Birmingham, England

Plastics in packaging

The last decade has seen a vast change in the packaging of common goods. Up to the mid-1950s, the predominant packaging materials were paper and cardboard for solids and tinplate and glass for liquids. The only plastic in common use in the packaging field was cellophane which, due to its higher cost, was not in direct competition with paper as a general purpose wrapping material. The chemical inertness of the common plastics, particularly of the polyolefins and polystyrene was known from the time of their discovery and they immediately offered advantages over metals in the packaging of corrosive liquids such as sanitary fluids, bleaches, etc. The development of liquid household detergents gave the main impetus to the design of a flexible general purpose dispenser which was later to become popular too in the cosmetics industry. Low density polyethylene was the polymer which fitted the requirements of the detergent pack almost ideally. The polymer was cheap enough to compete with any other packaging materials and, because of its low softening temperature, it could be readily moulded. Like glass and metals it had excellent water barrier properties. Unlike glass and metals it was both flexible enough to permit distortion to eject liquids or powders by pressure and also tough enough to resist heavy impact. Its chemical resistance at ambient temperatures was as good as glass and much better than metals.

At this time no one gave very much thought to the question of disposal of plastic packaging. It was assumed that the standard methods of dealing with household refuse, namely tipping, would suffice. Over the years, however, it has become apparent that this disposal technique is not adequate for the newer packaging materials. Whereas glass breaks and has traditionally been used in returnable containers and paper, although discarded, rapidly biodegrades in the environment, plastics do neither of these. Low density polyethylene (LDPE), high density polyethylene (HDPE) and poly (vinyl chloride) (PVC) are not only very long lived in the environment, but they also maintain their physical properties for many years of outdoor exposure. After lying exposed to the weather for many years a typical LDPE package will still be intact and show no sign of general disintegration. The first packages to show any evidence of physical breakdown, starting on the upper surface exposed to the weather, will be those made from HDPE. This polymer is much more rigid then LDPE and is normally used for packaging of corrosive chemicals such as bleaches, sanitary fluids, pharmaceutical and veterinary products. The reason for the greater rigidity of HDPE over LDPE is its higher crystallinity and this is also the reason for its reduced toughness and faster breakdown in the environment.

Polypropylene (PP), which is not at present widely used in liquids packaging, is somewhat less resistant to environmental breakdown than is HDPE, and the more recently developed high impact polymers such as high impact polystyrene (HIPS), impact modified PVC and acrylonitrile–butadiene–styrene copolymers (ABS) are similar in behaviour. The impact modified polymers are rapidly expanding into this area of packaging due to their excellent physical properties which allow them to be used in very thin cross section.

A parallel development which has occurred over the past ten years is the replacement of cellulose based wrapping and packaging materials (paper, cardboard, wood, cellophane) by polyethylene (mainly low density) and polypropylene film. Perhaps the major change has occurred in agricultural packaging where LDPE has almost entirely replaced paper for sacks. It has been reported that almost 100,000 tons of LDPE are used annually for agricultural sacks in the UK alone.[1] Most of this plastic remains in the fields for many years since, like the detergent bottles made from this polymer, packaging film breaks down extremely slowly. Even after many years of outdoor exposure it does not embrittle cleanly but shreds under mechanical stress. There is quite a pronounced effect of film thickness with all film-forming plastics. Table 1.1 shows the effect of increasing sample thickness on embrittlement time in the case of polypropylene.

Table 1.1 *Outdoor lifetime of polypropylene as a function of sample thickness*

Nature of sample	Thickness (ins)	Irradiation dose to produce 50% loss of strength (Langleys)
Moulded bar	0.062	52 600
Monofil	0.005	33 600
Multifil	0.001	16 000

The colour of the plastic also has a marked effect on degradation rate. As little as 1 per cent of carbon black extends the life of LDPE film by a factor of twenty-five.[2] It is rather unfortunate therefore that black pigmented sacks are so widely used in refuse collection since not only are they themselves virtually indestructible in the environment but they also protect their contents from UV degradation.

Another recent development has been the replacement of hessian ropes and twines by the much cheaper polypropylene fibres. Industrial ropes and fishing nets are now almost entirely made from this cheaper fibre. Unlike hessian, polypropylene does not biodegrade in the environment and after discard due to mechanical wear it retains its physical properties for a considerable time. Vast quantities of polypropylene ropes and fishing nets are discarded at sea and because the polymer has a density less than that of water, the materials float to the seashore where they accumulate.

A new form of polypropylene fibre is much cheaper to fabricate than that currently used for ropes and fishing nets. This is made by stretching sheets of polypropylene and slitting the orientated film into fibre which closely resembles hessian. Because of its cheapness this is now being widely used in twines, particularly for packaging purposes, but it suffers from the same disadvantage already mentioned in that it does not disappear by biodegradation in the environment.

Effect of plastics on the environment: potential solutions

The interpretations of the word pollution range from a narrow definition in terms of a direct threat to life to the broad interference with the ecological balance. The Oxford Dictionary definition of pollute is: to make physically impure, foul or filthy; to dirty, stain, taint, befoul. It has been suggested that plastics do not, at present, constitute a pollution hazard in the first sense because they are chemically and biologically inert and cannot interfere with life.

Even on the basis of the narrower definition there is some question as to whether discarded plastics are harmless in the environment. Death of cattle and deer have been reported due to ingestion of plastic film. Within the broader description there can be no doubt that plastics waste is aesthetically distasteful, particularly in recreational areas such as beaches and in the countryside.

The use of plastics in packaging is bound to increase. Since 1962, the annual rate of increase of plastics in packaging has been 25 per cent.[1] It has been said that the percentage of plastics in domestic waste is growing in proportion to the number of supermarkets since the supermarket depends almost entirely on cheap packaging for its viability. The United States, which has had supermarkets longer than Britain, has a plastics pollution problem in a much more severe form. In the United Kingdom the amount of plastic in domestic refuse is less than 2 per cent,[3] whereas in certain parts of the United States and in Japan, it is approaching 10 per cent.

The problem is a serious one and must be tackled rationally and on the basis of sound scientific principles if there is to be minimum interference with commercial enterprise and individual liberty. Although legislation may be necessary eventually to control this threat to the environment, it should not be undertaken without a thorough evaluation of the facts as they relate to the natural environmental stability of existing packaging plastics, their present means of disposal and the possible development of this field in the future. A step in this direction has been the setting up in 1971, under the auspices of the World Packaging Organisation, of a Working Party to examine the effect of all the factors which determine packaging disposability (for example, materials of construction, package design, etc.) on possible disposal techniques. The disposal techniques currently used in the UK will be reviewed in the next section in the light of the increasing proportion of plastics in domestic and industrial waste.

Existing practice in waste disposal

The most important techniques which are currently used for the disposal of solid waste products are tipping and incineration.[3] These should not be looked upon as mutually exclusive since the economics of using one rather than the other depends to a very large extent upon the circumstances. Tipping is the favoured method in rural districts where there are large areas of waste ground or empty holes close at hand. If the waste contains an appreciable proportion of plastic, however, it cannot normally be used for building purposes due to the persistence of the long-lived plastics in the ground which make it unsuitable for foundations. Incineration is favoured in urban areas where transport costs would prohibit disposal by tipping. Incineration is the easiest to organise but involves considerable capital cost. Moreover capital cost will increase steadily with increasing quantities of corrosive gases in the effluent as is the case when the waste contains an appreciable proportion of PVC which undergoes decomposition, with evolution of hydrogen chloride gas.

$$-CH_2-CH-CH_2-CH-CH_2- \quad \longrightarrow \quad -CH{=}CH-CH{=}CH-CH{=}$$
$$\qquad\quad | \qquad\quad | \qquad\qquad\qquad\qquad\qquad +2HCl$$
$$\qquad\quad Cl \qquad\quad Cl$$

I

Both of the above methods must become decreasingly favoured as the amount of plastic waste in the discard increases.

Among the newer methods which are being considered is pyrolysis (that is, heating in the absence of air) which may give rise to gaseous and liquid fuels[3]; and in those cases where individual plastics can be separated, useful chemical intermediates can also be obtained. For example, polystyrene readily depolymerises to styrene monomer.

Recycling of plastics by reprocessing is, in principle, very attractive since it offers the possibility of converting waste plastic to useful articles without going through the chemical manufacturing stage. On the basis of present waste management technology it is economically out of the question for packaging waste in general but it offers promise where waste plastic can be collected in bulk relatively cheaply and in a form relatively uncontaminated by other waste products. This kind of recovery process is already used in plastics factories which produce a proportion of waste from the fabrication process, and it is clearly economic to return this directly to the compounding stage of the process. Immediately a separation process is involved, and particularly when the plastic waste has to be cleaned before feeding into the extrusion or injection moulding machine, then it is at present cheaper to start with raw polymer. A further problem associated with reprocessing is that very small amounts of minor contaminants introduced before the processing operation may have a profound effect on both the thermal and the subsequent service stability of the plastic. This therefore would normally necessitate the incorporation of relatively expensive antioxidants and stabilisers in such quantity that any economic advantage of starting with a cheap raw material would probably be lost. However, there may be a limited number of applications for domestic reprocessed waste plastics where short life and inferior colour are not a disadvantage. It could almost certainly never be applied to packaging materials for aesthetic and hygiene reasons.

It will be evident from what has been said so far that the disposal techniques discussed, whilst suitable at least in principle for domestic and industrial waste, where collection for incineration or other types of destruction is feasible, cannot be applied to sea-borne refuse or public litter or even to agricultural litter where the cost of collection would be prohibitive. The ideal approach in this situtation is to restore, as far as possible, the natural balance by chemically modifying the plastic so that it breaks down in the environment at a rate similar to the natural polymer, cellulose. Before discussing how this might be done, it is necessary to understand the difference in molecular structure between cellulose and the synthetic polymers.

Chemical and physical structure of polymers

Cellulose is a high molecular weight polymer containing the repeating unit II.

II

Owing to the existence of a high concentration of hydroxyl groups, the polymer contains strong hydrogen bonds and is highly ordered. It is consequently both very high melting and very strong. However, again due to the presence of hydroxyl groups, cellulose can absorb a considerable proportion of water into its structure and it is the breaking and reforming of the hydrogen bonds between hydroxyl groups in the chains in the presence of water which lead to the well known tendency of cotton to 'wrinkle' on drying. It is also the hydrophilic nature of cellulose that permits its rapid biodegradation by the absorption of water-borne micro-organisms, and this is the reason for the rapid disappearance of finely divided wood pulp (paper) in the environment. Cellulose is not readily oxidised by molecular oxygen although the presence of certain dyestuffs can lead to its accelerated degradation or tendering.[4]

The vinyl polymers used in packaging all contain the repeating unit III.

X	Polymer
—H	Polyethylene (PE)
—CH₃	Polypropylene (PP)
—Cl	Polyvinylchloride (PVC)
—⬡	Polystyrene (PS)

III

Since they do not contain a hydrophilic group, such as hydroxyl, they are not swollen by water in their normal state and it is this property which gives rise to their valuable barrier properties as packaging materials. By the same token, however, they are not attacked by water-borne bacteria normally present in the environment. Pure vinyl polymers are, nevertheless, very subject to attack by atmospheric oxygen, particularly at elevated temperatures during fabrication or on exposure to the environment.[4] Thus, for example, if unstabilised polypropylene is heated on an open mill at 165 °C it changes rapidly within a few minutes from a viscous polymer melt to a mobile liquid due to a catastrophic reduction in molecular weight. The other vinyl polymers show a similar behaviour in their pure state to a greater or lesser extent. The melt instability of the vinyl polymers as measured by an increase in the melt flow index of the polymer* (see figure 1.1) is clearly a disadvantage during their manufacture since injection moulding, blow moulding, etc., involve the melting of the polymer at high temperatures and it is impossible to exclude molecular oxygen completely. Consequently the

---

* Melt flow index is the amount of polymer extruded through a standard orifice in a given time. It is inversely related to molecular weight.

polymer scientist has developed melt stabilisers (antioxidant, AH) for vinyl
polymers (PH) which interfere with the oxidation chain reaction by providing
a more energetically favourable route for the reaction of alkylperoxy

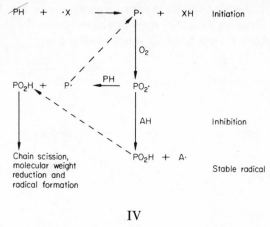

IV

radicals.[4] Since the latter are the radical species which predominate in the
chain reaction, this leads to inhibition of the overall process for as long as the
antioxidant survives in the polymer (see IV above).

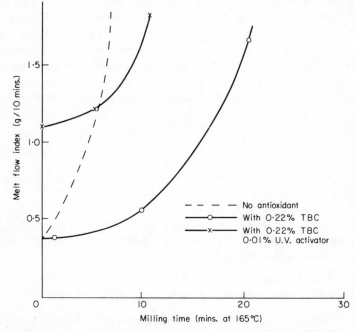

Figure 1.1: Change of polypropylene Melt Flow Index with milling time

Physical breakdown of the polymer is a direct consequence of the reduction in molecular weight of the polymer which is associated with the breaking of the polymer chain. This is the physical consequence of the chemical breakdown of the highly unstable polymer hydroperoxide PO_2H at elevated temperatures. Thus, for example, if PH is polyethylene, the derived hydroperoxide (PO_2H) is known to break down as shown in V.

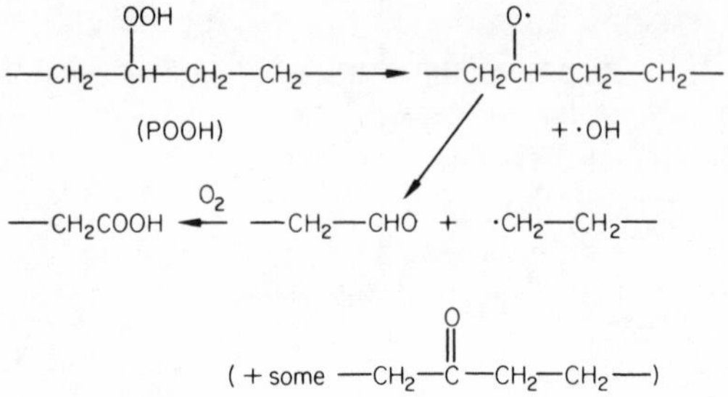

V

This chain scission process itself produces free radicals which can continue the kinetic chain reaction and thus give rise to the well known phenomenon of the 'branching chain reaction' of which the autoaccelerating rate curve shown in figure 1.1 is a direct consequence. The reasons for the increasing rate of oxidation with time will be evident from the oxidation scheme shown (V). Initially there is little or no hydroperoxide present in the polymer and hence the initiation process will depend on the formation of an unspecified free radical X· in a very slow process. This may be, for example, the energetically unfavourable attack of oxygen on the polymer as shown in VI.

$$PH \; + \; O_2 \longrightarrow P\cdot \; + \; \cdot O_2H$$

VI

However, as hydroperoxide builds up so it will act as its own initiator and it has been shown that in the early stages of cyclohexene autoxidation, the rate of oxidation is proportional to the concentration of hydroperoxide which has formed in the system.[5]

Sensitisation of synthetic polymers to environmental degradation

It will be evident from the above discussion that several ways of sensitising polymers to the environment are possible.

The first approach would be to attempt to increase the rate of attack of environmental micro-organisms. This can be approached in two ways, either by the development of new packaging polymers by chemical modification at the manufacturing stage which will be susceptible to attack by existing bacteria normally present in the environment, or by the culture of new strains of micro-organism to attack the present packaging polymers which are inert in bulk form to existing bacteria. Both of these have been discussed as possibilities but so far no useful developments have occurred.[6,7] Moreover, both approaches have inherent weaknesses.

As was discussed earlier, the development of the thermoplastics as packaging materials resulted, at least in part, from their good barrier properties. Any structural modification which leads to water swelling or even increased rate of moisture transport must seriously restrict their usefulness. A second advantage is their extreme cheapness which results largely from the simplicity of the chemical building units and of the processes involved. This has in turn led to their manufacture on a very large scale for a wide range of applications besides packaging. The manufacture of a new polymer would amost inevitably be more expensive and would have to be shown superior to be viable. It has recently been reported[8] that a cellulose-based water-soluble polymer has been developed for bottles which is protected by a double skin of water impermeable thermoplastic which can be punctured before disposal of the bottle. Although the cellulose material will dissolve and biodegrade rapidly these plastics are considerably more expensive than existing materials and since the skins are not destroyed, the problem is only partially solved.

The second approach seems to be of doubtful value due to its potential lack of control. If bacteria could be developed which could attack any polymer article from the outside, it is difficult to see how their attack could be limited to those plastics which were being used only in short-lived applications.

However, a useful preliminary study was carried out by Jen-Hao and Schwartz,[9] who found that bacteria which grew well on low molecular weight paraffins were able to attack polyethylene to the extent of removing the low molecular weight polymer but were unable to degrade the high molecular weight material. This has relevance to the bio-degradative breakdown of plastics which have already been reduced to low molecular weight material by some other process (see below).

As has already been discussed at least some of the common packaging plastics are sensitive to the environment and break down over a long period of time to give fine particles which appear to be absorbed into the environment. Examination of the ground surrounding such samples has failed to show the presence of any plastic particles. This accords with studies in other chemical compounds in which it has been shown that the rate of biological degradation is related to the degree of particle subdivision.[4]

Even more important, however, is the chemical modification which occurs in plastics when they break down by oxidation. The types of chemical groups which are ultimately formed by breakdown of hydroperoxides (IV) are carbonyl, hydroxyl, carboxyl and related polar groups which can readily form hydrogen bonds with water. This will lead to a greater degree of hydrophilicity than in the unmodified polymer which will allow the more ready attack of water-borne micro-organisms. In accord with this view, it is known that poly (vinyl acetate) and melamine-formaldehyde resins, both of which contain carbonyl and other polar groups, are more susceptible to biological attack even in massive form than are the hydrophobic packaging materials.[10,11]

It seems likely then that the approach to biodegradation through accelerated oxidative degradation of polymers in the environment offers some prospect of limiting the lifetime of discarded packages out-of-doors and a number of university research groups have recently reported success in this approach.

Photochemistry of polymers

An interesting and logical approach to this problem is that of Guillet and his co-workers[12,13] who made use of the extensive fundamental research investigations of Norrish and his research school on the photo-chemistry of the carbonyl group.

It was known that when ketones are irradiated, the first short-lived product formed is the excited singlet which rapidly reverts to the ground state with fluorescent emission or undergoes the electronically 'forbidden' inter-system crossing to the triplet state. From the chemical point of view, the highly reactive singlet state is too short-lived to participate extensively in chemical reactions but the longer-lived triplet state has a much higher probability of interacting before reverting to the ground state by phosphorescent emission.

The electrons in the triplet state are unpaired and its reactions are therefore analogous to the reactions of highly reactive free radicals. Their formation in an autoxidising system will therefore normally lead to rapid initiation of autoxidation and it has been suggested[14] that the small amount of carbonyl present in polymer chains is responsible for their photosensitivity.

Hydrogen abstraction and other radical type reactions is only one of the possible ways in which triplet carbonyl can react although this predominates in aromatic ketones[15,16] (for example benzophenone). In this case the triplet hydrogen abstracts from the solvent giving alkyl radicals which can then participate in autoxidation. Derivatives of tetraphenyl-ethane are formed by

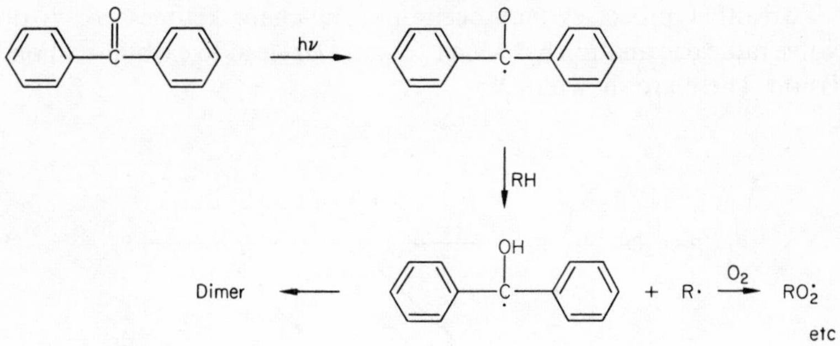

VII

dimerisation of the resonance stabilised diphenyl hydroxymethyl radical (VIII).

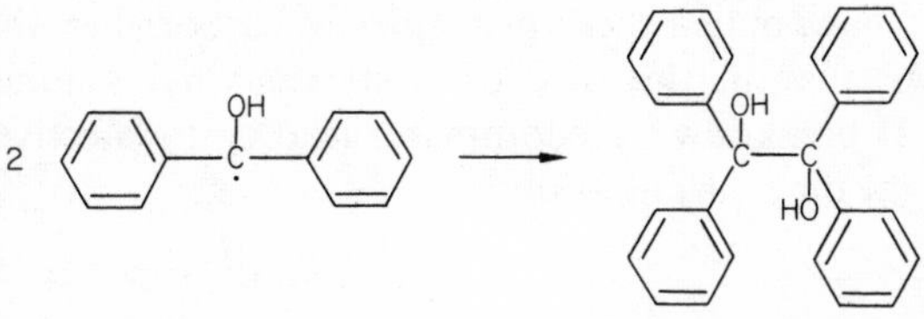

VIII

In the case of aliphatic triplet carbonyl more complicated reactions may occur which may or may not lead to radicals in the system. The initially formed triplet is in this case much more highly reactive and the radicals formed are shown in IX.

Reaction (a) is the Norrish type I breakdown of the carbonyl triplet.

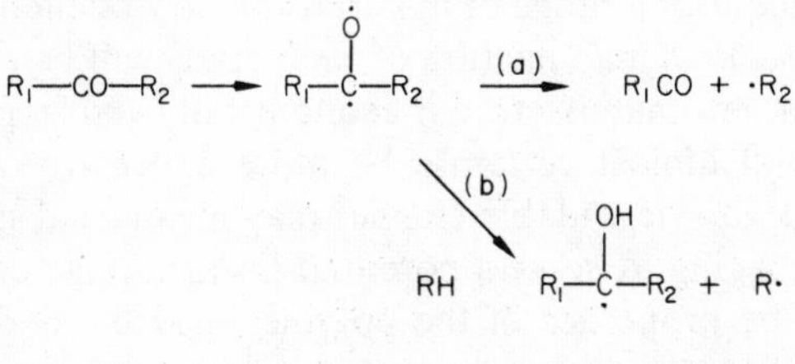

IX

Two other processes may occur in long-chain ketones which do not involve radical formation but which nevertheless lead to carbon-carbon bond scission. These are shown in X.

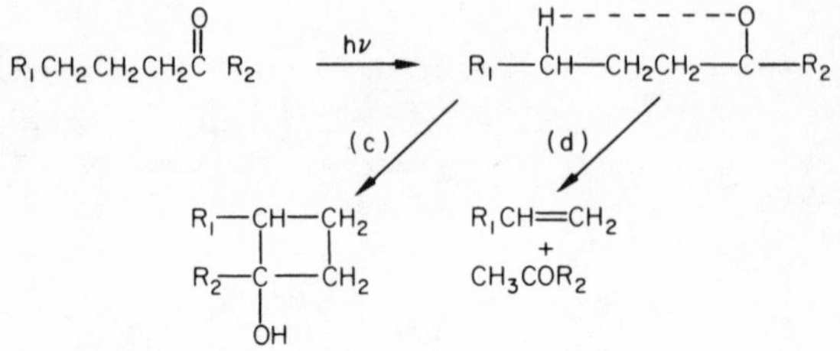

X

Guillet[12,13] has applied these principles to carbonyl modified polymers and has shown that in the case of a styrene-vinyl ketone co-polymer, Norrish Type II breakdown predominates leading to effective reduction in molecular weight on UV irradiation.[13]

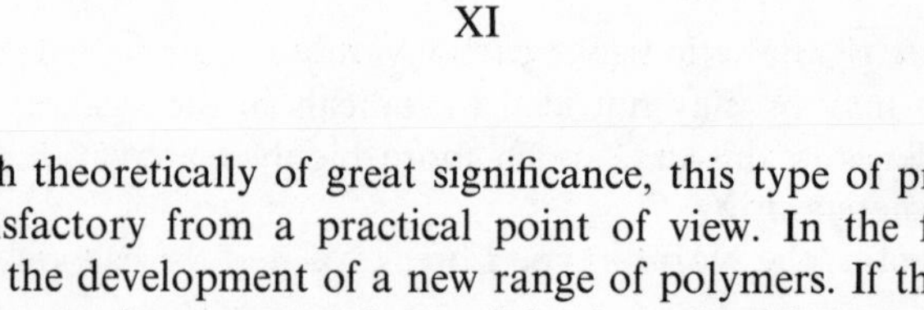

XI

Although theoretically of great significance, this type of process is not entirely satisfactory from a practical point of view. In the first place it necessitates the development of a new range of polymers. If the packaging manufacturer requires a range of materials of varying lifetime then this may mean that the scale of manufacture of each grade will be reduced below the level at which a new manufacture is economically viable, particularly as the co-monomer will almost certainly be more expensive. Alternatively, the production of photo-degradable plastics may mean a substantial increase in the cost of packaging. A second potential disadvantage of polymer modification is that the properties of the polymer may be modified particularly with respect to water transport properties, which would involve an expensive assessment of the suitability of the new polymers for packaging purposes.

A different approach has been adopted by other workers which avoids some of the difficulties encountered in modifying the polymer at the manufacturing stage. This is to incorporate additives during the conversion of polymer to the finished product which catalyse the oxidative processes occurring in the presence of UV light. The principles involved have already been discussed in relation to the transformation of electromagnetic energy into chemical energy by the carbonyl groups. A number of chromophores are known to be capable of doing this and indeed the 'tendering' effect of dyestuffs on cotton by visible light is known to be due to a photochemical process since the dyestuffs are themselves destroyed during the course of irradiation.[4] The pre-requisites for an effective UV activation on packaging plastics have been outlined by Scott.[17] These are that the additive should not interfere with the processing operation, the plastic should not be affected by diffuse daylight, only by the shorter wavelengths of the sun's spectrum normally found outdoors, the outdoor lifetime should be predictable and variable by alteration of the additive concentration and finally that there must be an adequate built-in warning system (for example, a change in colour which parallels the rate of degradation of the plastic).

Some of the substituted benzophenones without 2-hydroxy groups appear to meet these requirements.[18] At concentrations between 1 and 2 per cent a rapid initial rate of photo-oxidation of polyethylene is achieved as measured by increase in carbonyl content of the polymer (see figure 1.2). This is paralleled by a rapid increase in modulus of the polymer with loss of tear strength. Unfortunately this process is autoretarding and although embrittlement eventually occurs, the formation of the stable diphenylhydroxymethyl

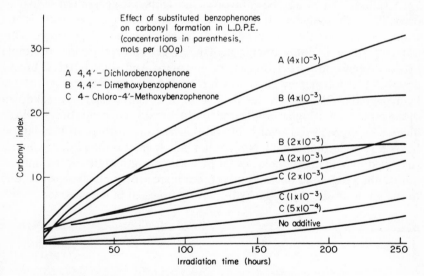

Figure 1.2: Effect of substituted benzophenones on carbonyl formation in LDPE

radical delays this process. It is interesting from the point of view of chemical mechanism that benzophenones containing a hydroxy group ortho to the carbonyl group are powerful UV stabilisers since, like the unsubstituted benzophenones, they have the ability to absorb UV light but this does not lead to the chemically active triplet state in this case since the 2-hydroxyl group appears to have the ability to internally deactivate this energetic species.[19]

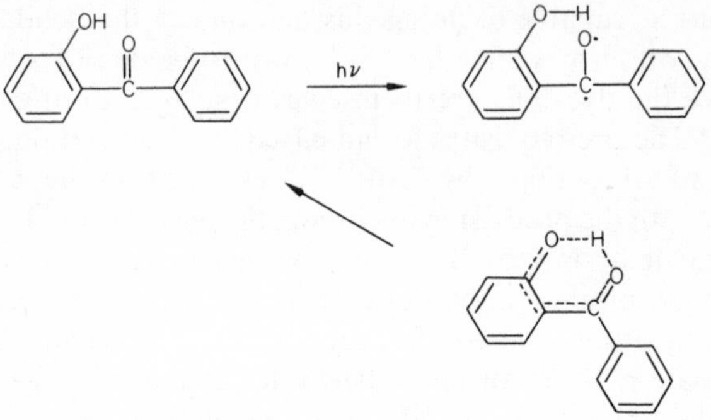

XII

There is some physical evidence for this process in that the most effective UV stabilisers for polymers are those which form the strongest hydrogen bond between the carbonyl and hydroxyl groups.

A rather different approach involves the deliberate generation of hydroperoxide in the polymer. It was shown earlier that hydroperoxide is the main, and initially the sole, product of the autoxidation of polymers. This is the reason for the autoacceleration of the oxygen absorption curve of simple hydrocarbons and pure hydrocarbon polymers. And since physical breakdown of the polymer by chain scission occurs through breakdown of the hydroperoxide (IV), the greater the degree of thermo-oxidative breakdown of the polymer, the higher will be the hydroperoxide concentration until the linear rate is reached at which the rate of peroxide formation is balanced by its rate of decomposition. Not surprisingly then, it is found that polymer subjected to oxidative conditions for increasing length of time (for example, during the processing operation) becomes increasingly photo-sensitive due to the photo-instability of the hydroperoxide present in the polymer.

$$ROOH \longrightarrow RO\cdot + \cdot OH$$

XIII

Figure 1.1 shows the effect of time of processing on a lightly stabilised polypropylene. After a slow start, the melt flow index of the polymer (which is inversely related to melt viscosity) increases rapidly. Figure 1.3 shows the effect of processing time on the growth of carbonyl in the polymer after

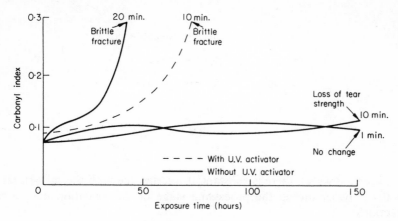

Figure 1.3:   Effect of milling time on the UV stability of polypropylene

compression moulding to film. It is clear that not only does the processing operation sensitise the polymer to oxidation but that this is accompanied by chain scission as would be expected if the photo-sensitive group were hydroperoxide. Unfortunately this approach, although mechanistically of great importance, is of limited usefulness to the polymer converter since it is essential that the melt-flow index of this polymer remains sensibly constant during the high temperature processing operation. Essentially what is required is an additive which will remain inert during the processing operation but which will catalyse the rapid formation of hydroperoxides subsequently. Carboxylic acid salts of transition metal ions are known to be effective photoactivators for vinyl polymerisation on irradiation and incorporation of ferric stearate into polypropylene was found to produce a similar catalysis of photo-oxidation (see figure 1.1, curve marked 'with photo-activators').[20] The reaction involved is almost certainly the generation of both acetoxy and alkyl radicals, since there is a rapid disappearance of the carboxylate absorption in the infra-red. The radicals produced are, of course,

$$(C_{17}H_{35}COO)_3 Fe \xrightarrow{h\nu} C_{17}H_{35}COO\cdot \quad + \quad (C_{17}H_{35}COO)_2 Fe$$
$$C_{17}H_{35} \quad + \quad CO_2$$

XIV

responsible for the formation of hydroperoxide by reaction with either the polymer or with oxygen.

$$C_{17}H_{35}COO\cdot \quad + \quad RH \longrightarrow C_{17}H_{35}COOH \quad + \quad R\cdot$$

$$\text{and } C_{17}H_{35}\cdot \quad + \quad O_2 \longrightarrow C_{17}H_{35}OO\cdot$$

$$C_{17}H_{35}OO\cdot \quad + \quad RH \longrightarrow C_{17}H_{35}OOH \quad + \quad R\cdot$$

XV

Again, however, there is substantial interference with the melt-stability of the polymer during the processing stage of the oxidation due to the reactions

$$ROOH \quad + \quad Fe^{3+} \longrightarrow RO\dot{O} \quad + \quad Fe^{2+} \quad + \quad H^+$$

$$ROOH \quad + \quad Fe^{2+} \longrightarrow RO\cdot \quad + \quad Fe^{3+} \quad + \quad OH^-$$

XVI

which lead to a catalysis of the oxidation process through the injection of free radicals.

A recent approach which has proved much more successful is the use of a transition metal ion such as ferric stearate in combination with a 'restraining agent' which prevents the participation of the ion in the above redox reactions. Indeed some combinations of transition metal ion and ligand, particularly when the ligand contains sulphur, have been found to exert a positive antioxidant effect in the processing operation. Thus, for example, ferric stearate and the complexing agent/antioxidant, zinc dibutyldithiocarbamate, stabilise polypropylene effectively at concentrations between 0.01 and 0.1 per cent but are effective UV activators for the polymer.[16] The preformed ferric dibutyl dithiocarbamate is equally effective. It appears likely that the antioxidant function is due to the peroxide-decomposing activity of the metal dithiocarbamate since it has been shown that these are powerful hydroperoxide decomposers in model systems. The evidence suggests that under autoxidation conditions they oxidise to give SO_2 which is a powerful catalyst for the ionic decomposition of hydroperoxides (see XVII).

$$(Bu_2NCSS)_3M \xrightarrow{ROOH} BuNCS + M_2(SO_4)_3 + SO_2$$

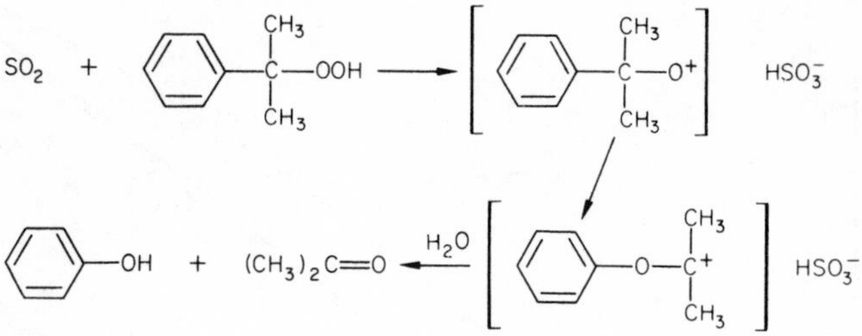

$$SO_2 + \underset{}{\bigcirc}\!\!-\!\!\overset{CH_3}{\underset{CH_3}{\overset{|}{C}}}\!\!-\!\!OOH \longrightarrow \left[\underset{}{\bigcirc}\!\!-\!\!\overset{CH_3}{\underset{CH_3}{\overset{|}{C}}}\!\!-\!\!O^+\right] \quad HSO_3^-$$

$$\underset{}{\bigcirc}\!\!-\!\!OH + (CH_3)_2C{=}O \xleftarrow{H_2O} \left[\underset{}{\bigcirc}\!\!-\!\!O\!\!-\!\!\overset{CH_3}{\underset{CH_3}{\overset{|}{C^+}}}\right] \quad HSO_3^-$$

XVII

Under conditions of irradiation, however, ferric dibutyldithiocarbamate is rapidly photolysed as shown by the disappearance over a period of only several hours of the associated UV absorption bands at 262 and 310 nm[16] and the photolysis is almost certainly responsible in part for the rapid pro-oxidant process which follows since thiyl radicals are known to be effective hydrogen abstracting agents.

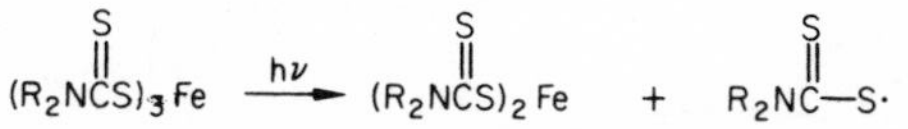

$$(R_2N\overset{S}{\overset{\|}{C}}S)_3Fe \xrightarrow{h\nu} (R_2N\overset{S}{\overset{\|}{C}}S)_2Fe + R_2N\overset{S}{\overset{\|}{C}}{-}S\cdot$$

XVIII

The detailed chemistry of this type of activation process is not yet clear, but it seems likely that once hydroperoxide is formed, this becomes the main initiator and is also involved in redox reactions with the transition metal ion. The effect of ferric dinonyldithiocarbamate on carbonyl formation in polyethylene is shown in figure 1.4. It is clear that the carbonyl concentration during initial photodegradation is lower in the UV activated system than it is in the control sample with no additive. Clearly then, carbonyl cannot be the UV activator in this case. The effect of 0·05 per cent of ferric dibutyldithiocarbamate on the elongation at break of LDPE film on outdoor exposure in England is shown in figure 1.5, and a similar curve for the change in tenacity of crystalline PP—orientated tape is shown in figure 1.6. A notable feature about all the degradations is that initially the polymer films containing the additives show a similar behaviour to the control without

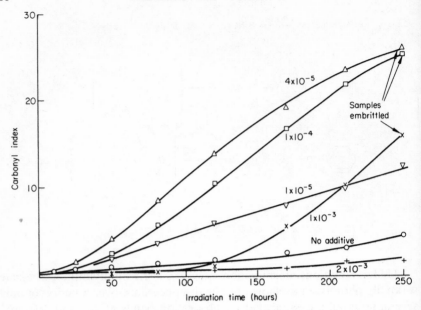

Figure 1.4:   Effect of Fe 111 D9DC on carbonyl formation in LDPE (numbers
              on curves are moles per 100 g of polymer)

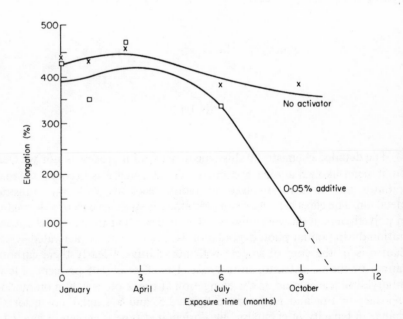

Figure 1.5:   Exposure of LDPE blown film (60 m) in England. Change in
              elongation in machine direction

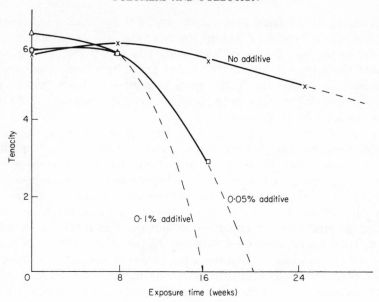

Figure 1.6: Exposure of polypropylene tapes in England. Change in tenacity

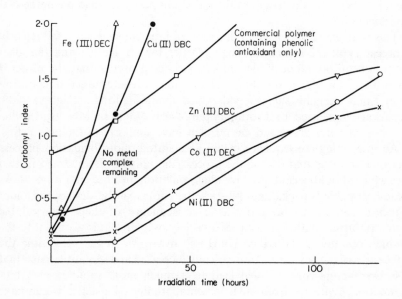

Figure 1.7: Effect of metal dithiocarbamates on carbonyl content of polyethylene during processing and accelerated UV exposure. Concentration of
metal complex = 2×10^{-3} moles/100 g

additive and then become auto-accelerating. This is the kind of behaviour which would be expected if hydroperoxide was initially absent and built up rapidly in the polymer. A detailed study has shown that all the metal dithiocarbamates but notably the nickel and cobalt compounds are UV stabilisers for polyolefins. They appear to exert their effect primarily by virtue of their ability to destroy hydroperoxides and hence remove the main photo-initiator for photo-oxidation.

Figure 1.7 shows the change in carbonyl index for severely processed polyethylene containing several metal dithiocarbamates at 1 per cent initial concentration in the polymer. It is clear that they have protected the polymer from thermal oxidative breakdown during the processing operation since the carbonyl index of the control with a conventional phenolic antioxidant is very much higher. The iron and copper complexes are essentially destroyed during the processing operation and do not show an initial UV stabilising effect. These agents have been shown to be powerful catalysts for hydroperoxide decomposition[21] and this effect clearly persists in the polymer on exposure to ultra-violet light until the metal complex is destroyed. A similar effect has been found for the other transition metal complexes of the dithiocarbamic acids. Although these are less stable under thermal-oxidative conditions, they do show an initial photostabilising effect which is directly related to the concentration of the metal complex. By controlling the concentration of both photoactivating and stabilising additives it is therefore possible to regulate the length of the induction period to match commercial requirements.

This is an effect of considerable practical importance since it will clearly be necessary in the future to design plastics for service life and the ideal situation from the point of view of the packaging user is that the physical properties of the plastic should not change for a predetermined time during use and that after this safe interval, which should be long enough to allow the package to be used for its primary purpose, it should rapidly disintegrate. High impact polystyrene and ABS resins behave similarly to the polyolefins.

An interesting consequence of the accelerated photo-oxidation process discussed above is that during the process the polymer is modified in two ways which should permit more rapid assimilation into the environment. As already discussed packaging plastics are not readily destroyed under biological conditions, primarily as a consequence of their hydrophobic nature. Photo-oxidation introduces a variety of hydrophilic groups into the polymer, notably, carbonyl, carboxyl and hydroxyl. At the same time the surface area of the polymer is dramatically increased and it is anticipated that these two factors together will lead to a much more rapid rate of biodegradation. Evidence from environmental studies on plastics which have been artificially photodegraded indicates that this is so but that very considerable reduction in molecular weight has to occur before this reaction becomes important.

References

1. *European Packaging Digest*, 1971, **91**, 2
2. W. L. Hawkins and F. H. Winslow (1965). *Reinforcement of Elastomers*, Interscience, N.Y., p. 563
3. J. J. P. Staudinger. *The Disposal of Plastics Waste and Litter*, S.C.I.— Monograph No. 35
4. G. Scott (1965). *Atmospheric Oxidation and Antioxidants*, Elsevier, Chapter 7
5. J. L. Bolland (1949). *Quart. Rev.*, **3**, 1
6. J. A. Fendley (1971). Plastics Institute Conference Supplement No. 4, p. 27
7. F. Rodriguez (1971). *Chem. Tech.*, July, 409
8. Anon (1971). *Business Week*, Nov. 13
9. L. Jen.-Hao and A. Schwartz (1961). *Kunststoffe*, **71** (6), 317
10. A. E. Brown (1944). *Modern Plastics*, **23**, 189
11. E. Abrams, Nat. Bur. Stand., Misc. Pubn. No. 188
12. G. H. Hartley and J. E. Guillet (1968). *Macromolecules*, **1**, 165
13. F. J. Golemba and J. E. Guillet (1970). *S.P.E. Journal*, p. 26
14. A. R. Burgess (1953). *Polymer Degrad. Mechanisms*, N.B.S. Circular, **525**, 149
15. G. S. Hammond and W. M. Moore (1959). *J. Am. Chem. Soc.*, **81**, 6334
16. G. O. Schenck (1957). Atti del 2° Cong. Int. Fotobiol., p. 29
17. G. Scott (1970). *Plastics, Rubbers, Textiles*, **1**, 361
18. G. Scott (1971). Plastics Institute, Plastics and Polymers Conference Supplement No. 4 (Sept.), 29
19. G. Scott (1965). *Atmospheric Oxidation and Antioxidants*, Elserier, Chapter 7, 180 ff
20. D. Mellor, A. Moir and G. Scott (1973). *Europ. Polym. J.*, **9**, 219
21. J. D. Holdsworth, G. Scott and D. Williams (1964). *J. Chem. Soc.*, p. 4692

CHAPTER 2

DOMESTIC REFUSE

Its composition, properties, recovery potential and disposal methods: present and future

A. PORTEOUS

Reader in Engineering Mechanics, The Open University, Walton Hall, Milton Keynes, Bucks

This chapter deals with refuse which falls within the local authority domain, that is, domestic refuse. This is refuse which is produced from households and hotels. The material is covered under the Public Health Act of 1936 and must be removed free of charge (some charge is usually made for bulky materials such as old refrigerators). Trades waste which emanates mainly from shops, offices and workplaces is often collected by local authorities but at a charge. Industrial and builders wastes are usually disposed of independently to councils tips or disposal plants for disposal on payment.

The quantity of domestic refuse now produced is circa 17,000,000 metric tons per year. This amount is increasing annually by around 1 per cent but the density is undergoing dramatic changes as the nature of the refuse alters in accordance with the current throw-away life style.

In discussing refuse disposal processes the prime input is the refuse composition.

Refuse composition

Domestic refuse is changing rapidly; the trend is to more paper, plastics and vegetable matter and less cinder ashes and glass.

Figure 2.1 shows the variation in refuse composition in the period 1953–69 for a large city. It is immediately obvious that paper is by far the largest individual component. This is a trend which has accelerated in recent years due to the introduction of Smoke Control Zones and the changes in living habits as the effects of affluence and urban reconstruction schemes are felt. Table 2.1 gives a qualitative seasonal analysis of household refuse for the City of Birmingham in 1969.[1] It is seen from this table that the paper content

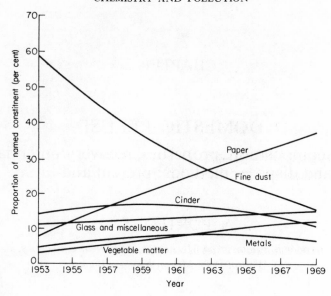

Figure 2.1:   Postwar variation of UK refuse characteristics for a city with a
population of 500,000 (Analysis)

averages 48 per cent by weight. The Birmingham refuse analysis differs
slightly from that of the UK as a whole due to the extensive Smoke Control
Areas and urban redevelopment which have taken place—the general trend
of higher paper and vegetable contents are foreshadowed by these figures.

Table 2.1 *Qualitative and quantitative seasonal analysis of domestic refuse—1969. City of
Birmingham*

	Average of artisan, middle class, and residential			
	Winter	Spring	Summer	Autumn
	%	%	%	%
Fine dust and small cinder (under $\frac{1}{2}''$)	19.80	19.36	7.70	16.93
Cinder content (between $\frac{1}{2}''$ and $1\frac{3}{4}''$)	2.40	1.23	0.35	1.67
Vegetable and putrecible content	11.00	9.77	16.24	17.78
Paper content	49.03	47.84	48.93	45.71
Metal content (ferrous and non-ferrous)	4.97	5.97	9.26	6.28
Rag content (including bagging and all textiles)	2.80	1.79	2.85	2.21
Glass content (bottles, jars and cullet)	6.77	6.75	11.30	6.94
Plastics	1.46	0.93	1.08	0.77
Unclassified debris (not classified above)	1.77	5.36	2.29	1.71

What is already evident from both table 2.1 and figure 2.1 is that the
paper and vegetable group together form the largest constituent of domestic

refuse and thus any disposal process has and will increasingly have to deal primarily with these materials. Paper salvaging has currently little effect on the paper content of refuse due to the expense of sorting and the inability of the trade to absorb the resulting low grade pulp if all the paper content were were salvaged and pulped. The economics of separate paper collection, with its associated sorting and baling have discouraged many local authorities from attempting this exercise, although this trend may be reversed in time.

We live on a planet whose resources are finite and which are being used up at ever increasing rates, for example oil reserves and copper may be significantly depleted by the year 2000. These are examples of non-renewable resources. There are also the renewable ones such as trees and plant life which store solar energy in the form of tissues and sugars and it is to those resources that man must increasingly turn. With the drying up of easily obtained materials and increases in cost there is an increasing emphasis on recycling of wastes of all kinds and refuse is no exception. The concept of recycling must of course be tempered by a realisation that we have a market place economy and thus the return from recycled components of refuse (or most other wastes) must be such that a net reduction in their overall disposal cost is obtained. This approach is not necessarily in keeping with the preservation of the environment but as long as economic guidelines are used for evaluation then this approach will prevail.

With the goal of 'total recycle' of refuse constituents, it is appropriate to examine the major components in detail and discuss their potential for the recovery of useful products.

Product recovery potential of domestic refuse

The principal constituent of refuse is cellulose, which is the major component of the paper and vegetable matter. Any refuse destruction or recovery process will probably have cellulose as its main feedstock, thus an in-depth look at its properties is required.

Cellulose is a carbohydrate polymer composed of long chains of glucose units connected at hydroxyl groups as shown in figure 2.2 and has the empirical formula $(C_6H_{10}O_5)_n$. It is the chief structural element and major constituent of the cell walls of trees and higher plants, cotton for instance is nearly pure cellulose. It is insoluble in water and tasteless and is a non-reducing carbohydrate. These properties are mainly due to its high molecular weight. Many important derivatives stem from it such as cellulose nitrate, cellulose acetate and rayon. For the purpose of this chapter the most important processes it can undergo are bacterial decomposition, hydrolysis, pyrolysis and combustion.

Bacteriological decomposition can be carried out by mesophilic or thermophilic bacteria (actinomycetes and fungi). The former will take literally aeons of time, the latter can decompose the bulk of the cellulose in under

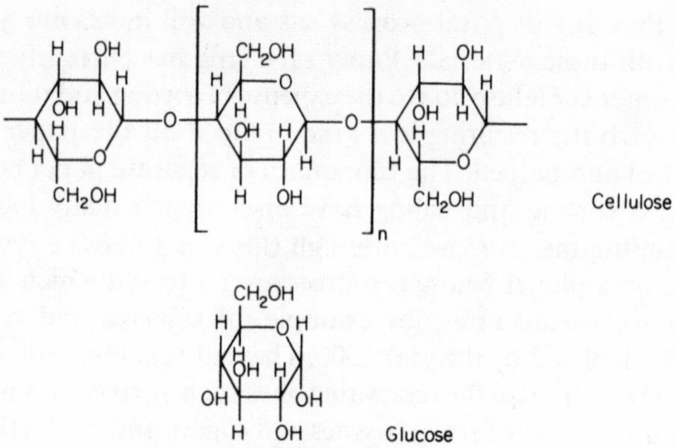

Figure 2.2:   Cellulose structure

four weeks given the proper conditions. From the foregoing, it is inferred that relatively rapid composting is possible—whether the compost is marketable is another matter.

Hydrolysis may be defined as a process in which a double decomposition reaction is carried out with water as one of the reactants. The hydrolysis of starch is the basis of the corn syrup industry. The hydrolysis of cellulose is accomplished with the aid of an acid as the hydrolysis reagent, as in I.

$$C_6H_{10}O_5 + H_2O \xrightarrow[H_2O]{H_2SO_4} C_6H_{12}O_6$$

$$\text{Cellulose} \qquad\qquad\qquad\qquad \text{Sugars}$$

I

The acid acts as a catalyst and for fast industrial reactions and high yields the process is carried out at elevated temperatures.

The cellulose to sugar conversion in I is a rate-controlled series reaction as the sugars decompose on continued exposure to the hydrolysing conditions. There are thus optimum reaction conditions for maximum sugar output as discussed by Porteous.[2]

Once the sugars are obtained, a variety of operations can be carried out, a principal one being fermentation to yield an aqueous solution of ethanol which is subsequently rectified. The fermentation reaction is given below (II).

$$C_6H_{12}O_6 \xrightarrow[\text{Nutrients}]{\text{Yeast}} 2C_2H_5OH + 2CO_2$$

$$\text{Sugars} \qquad\qquad\qquad \text{Ethanol}$$

II

Pyrolysis or destructive distillation is carried out by roasting the cellulosic material in retorts at high temperature and results in the following products and yields given in table 2.2.

Table 2.2 *Pyrolysis of Cellulose*

Product	% by weight
Charcoal	25–37
Water of pyrolysis	34
Acetone	0.3
Acetic acid	3.1
Tar and oils	10–15
Combustible gases	6– 8
CO_2	10–12

Tars begin to come off at 200 °C and a slightly exothermic reaction takes place at about 270 °C due to the occurrence of secondary and tertiary reactions. Further elevation of the temperature results in a slow formation of products, and no further volatilisation occurs at 470 °C.

Pyrolysis of refuse will be discussed in detail later in the chapter; the oil and gas yields are a function of the refuse composition and pyrolysis conditions. This process looks promising as it means that a replenishable energy source may become available which could augment diminishing fossil fuel reserves.

For combustion purposes the gross calorific value of cellulose is 17,400 kJ/kg (7,500 Btu/lb) whereas the average gross calorific value of mixed refuse is around 10,400 kJ/kg (4,500 Btu/lb) due to the ash, cinder and vegetable components. In the US where higher refuse paper contents are normal the gross calorific value is roughly 11,600 kJ/kg (5,000 Btu/lb). Thus the cellulose (paper) content of refuse has a marked influence on incinerator design. With gross calorific values below 6,950 kJ/kg (3,000 Btu/lb) supplementary fuel is usually required to ensure continuous and complete combustion.

Cellulose (together with any fats) constitutes the organic portion of refuse whereas glass, ashes and tin cans constitute the inorganic portion and brief mention should be made of them from a recovery point of view.

The recovery of glass cullets is no longer an economical operation, the same can be said of the ash content due to its low calorific value and the value of these components is only as a fill material. Metal recovery (usually tins) is often practised by magnetic extraction and baling as a commercial grade scrap can be obtained. At the present time tin can baling covers the costs of recovery handsomely. On conservation and recycling grounds, a metals extracting operation in conjunction with other processes is to be commended and the US Bureau of Mines have been looking at recovery of metals from refuse as discussed by Dean et al.[3]

We can now look at various refuse disposal processes in detail and analyse both their disposal performance and recycling aspects (if any) respectively.

Controlled tipping or sanitary landfill

The dumping of refuse on land is a disposal method used from time immemorial. Even today well over 80 per cent of the UK refuse is disposed of by the infilling of clay pits, coastal wet lands, valleys and in fact virtually any hole or depression further than 2–300 metres from housing and within economical handling distance is a potential candidate for landfilling.

Controlled tipping is probably going to continue in dominance as a disposal method but may sometimes be little better than an open dump. In this chapter controlled tipping means an installation where a satisfactory nuisance-free refuse disposal operation is being carried out in accordance with strict control procedures—the refuse being dumped in layers, compacted and covered with inert material.

The first essential in proper tipping is site selection. The principal parameters are:

(1) Length of haul—this is usually a major factor in disposal cost estimates; however, with near at hand sites being exhausted for many cities hauls of up to 50 miles may be economical before other alternatives are competitive. If long hauls are contemplated the refuse trucks will discharge at a transfer point where special lorries or trains haul the refuse to the disposal site.

(2) Prevention of water pollution—the site should not contain static water as the decaying refuse and water combination produce extremely offensive smells and encourage the breeding of flies. Running water should not be present, neither should the site connect with the aquifer as contamination of water supplies can result and jeopardise public health. There is evidence that some badly selected crude refuse disposal sites have polluted water supplies in the UK and geological surveys should be undertaken, if necessary, to guard water resources.

Even if water abstraction or catchment is not practised in the vicinity of the tip, amenities should not be destroyed. The decaying organic content of the refuse can impose an oxygen demand on any water percolating through the tip or on a connecting stream and render it unfit for fish life.

The crucial role of oxygen in the biological oxidation of organic carbon and nitrogenous materials is given in equations III and IV below.

$$\left.\begin{array}{l}\text{Organic materials} + O_2 \xrightarrow{\text{Micro-organisms}} \\ \text{Micro-organisms} + CO_2 + H_2O + NH_3\end{array}\right\}$$

III

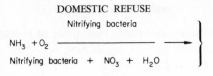

IV

The oxygen comes from the dissolved oxygen content of the stream or static body of water and is rapidly used up rendering it sterile and lifeless. Once the oxygen is consumed the decay processes become anaerobic and if, for example, sulphates caused by the refuse are present in the water, the bacteria will reduce them to sulphides which can react to form hydrogen sulphide (H_2S) with its associated foul odour. Thus any stream must be culverted, and any static water must be removed in a properly run controlled tipping operation. Consultation with River Authorities and water supply under-takings is normal practice before a new tip is begun and restrictions may be placed which will curb gross pollution.

(3) Distance from housing—the tip must be at least 200 metres from any housing to minimise nuisance.

(4) Availability of covering material—a properly run tip has the refuse in compacted layers not more than 1.8 m deep and these layers must be covered with at least 20 cm of inert material such as soil or ashes or thoroughly composted refuse. This prevents or controls the ingress of flies, rats, gulls and other forms of bird, insect and animal life. The covering of the refuse layers is an essential part of the operation and neglect of this aspect can cause the tip to degenerate to a major public nuisance.

(5) Wind direction—this cannot be controlled thus screens may be required to prevent airborne litter deposition.

Due consideration of the above factors can make controlled tipping a very acceptable method of refuse disposal, especially if the refuse is treated by pulverisation beforehand to aid compaction. This method of disposal will continue in dominance for many years as its costs are normally well below those of incineration or other disposal processes.

Controlled tipping on derelict ground or the infilling of depressions has provided many municipalities with excellent sites for playing fields and amenity areas. Where the refuse is not used for land reclamation but is merely dumped in old clay pits, etc., and provides no public amenity, it is the writer's opinion that this constitutes a waste of resources which should be recycled wherever it is practically possible.

Pulverisation

Pulverisation covers a variety of refuse treatment processes which result in the disintegration of the refuse from its crude state to a homogenously sized mix. Pulverisation is invariably used prior to composting of refuse and sometimes incineration, but in this section we shall discuss it from the controlled tipping viewpoint.

There is a strong body of opinion which states that crude refuse can never be tipped satisfactorily as compaction is difficult and the proper control of loose materials is often impossible. Furthermore, cover material in adequate quantities for the sealing of crude refuse tips is often just not available.

Pulverisation it is claimed gives a very homogenous material with a higher initial density than crude refuse ($904\,kg/m^3$ compared with $598\,kg/m^3$). Compaction is enhanced and less cover material is required. Furthermore land on which crude refuse cannot be tipped can often receive pulverised refuse easily without causing public nuisance.

The pulverisation process itself may be carried out in the wet or dry state. Wet machines are usually horizontally perforated rotating drums. The perforation size, speed and time in the drum control the particle size. Water is sprayed into the drum to assist the disintegration of the refuse.

Dry pulverisation is usually accomplished by hammer mills. The impact of the hammers effect the material reduction. The hammers may be free to swing on a rotating shaft or shafts or simply fixed to the shaft. Hammer mills give more accurate size control of the end product compared with the horizontal drum wet process.

A very successful pulverisation and landfill of coastal mud flats has been accomplished at Poole for land reclamation for road construction—there is no doubt that pulverisation can assist the management of tipping operations, but again the question of cost arises. However this form of pretreatment may increase as tip management standards rise.

Composting

Composting is a method of handling and processing municipal refuse which produces as end products (a) a humus-like material which may be used as a soil conditioner or top dressing for a controlled tip, and (b) rejected matter such as salvage and non-cumbustible material. Technically, composting is a biological process of decomposition carried out under controlled conditions of ventilation, temperature and moisture by organisms in the wastes themselves. Figure 2.3 shows the temperature time curve for a controlled composting cycle. The composting process falls into four readily identifiable stages as discussed by Gray and Biddlestone,[4] namely, mesophilic, thermophilic, cooling and maturing. The mesophilic organisms commence the decay process and die off when the temperature approaches 40 °C. Above 40 °C the thermophilic organisms (actinomycetes and fungi) take over, and the pH turns alkaline. Above 60 °C the thermophilic fungi die off but the actinomycetes can continue. The reaction degenerates when the biodegradables are consumed and the mass cools down. The thermophilic fungi then recommence and attack the cellulose which can take three or four weeks to be digested. Eventually activity dies off and the temperature approaches ambient and pH stabilises around 5, that is, slightly acidic.

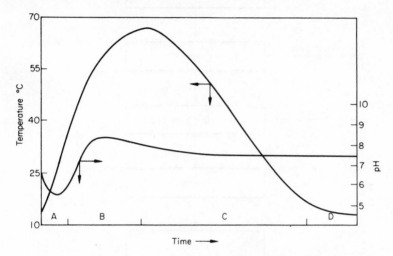

Figure 2.3: Temperature and pH versus time profile for compost manufacture (ref. 4)

The process rate is influenced by particle size, availability of nutrients, moisture, temperature and aeration. Refuse composting is characterised by extensive materials handling and separation. While many different methods have been developed there are certain common operations as outlined in the flow diagram, figure 2.4. Some of the operations shown are optional.

A standardised cost format has been developed[5] by which refuse disposal processes may be compared. For a highly automated compost plant, the capital cost is £4,500 per tonne of daily installed capacity and the disposal cost is estimated at £3·7 per tonne of input refuse—no credit was assumed for compost sales.

Possible sources of revenue cited for municipal compost are the proceeds from the sale of salvaged material and compost for use as a soil conditioner.

So far, salvaging of metals (with the exception of ferrous by magnets) and glass is not very economical as hand sorting is usually required. No market has yet been developed for the compost itself in the UK and there is strong reluctance on the part of municipal authorities to enter the marketing field with a high bulk, low selling price product. Some schemes have been quoted whereby compost is to be sold as an enriched fertiliser, but farmers have shown little interest in this development. The reluctance is due in part to the fact that basic compost from municipal refuse contains a very high carbon-nitrogen ratio of order 30:1, that is its fertiliser content is

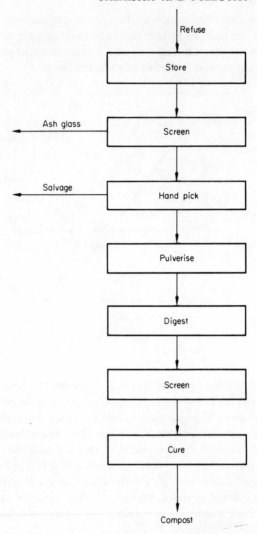

Figure 2.4: Flow diagram, compost from refuse

minimal. This is due to the composition of refuse as nitrogeneous compounds are almost non-existent. Thus all the compost does is provide a coarse form of humus unless enriched. It is noted that heavy metals can be present in refuse compost and may constitute a problem in its agricultural use.

After composting, 50–60 per cent of the original volume of refuse remains to be disposed of either by sale or tipping. By its very nature compost is not suitable for use as a stable fill for heavy building or road construction. It can, however, be spread in many localities where crude refuse cannot, even under controlled tipping conditions.

The cost of compost plants can vary widely, and an order of magnitude installed cost for a nominal 100 tonne per day digester plant complete with all materials, handling and grinding equipment is £450,000 of which roughly £250,000 is equipment costs.

Incineration

Incineration is the term used for the combustion of municipal refuse. In a properly designed and operated incinerator there is a substantial reduction in the volume of waste material to be disposed of by tipping.

The process is extremely hygienic and many of the problems associated with controlled tipping, such as windblown refuse, rodents and flies, etc., are completely eliminated. Properly incinerated refuse becomes a sterile ash with minimal carbon or fat content and thus can be safely tipped in almost any location. The greatest point in its favour is, of course, the substantial

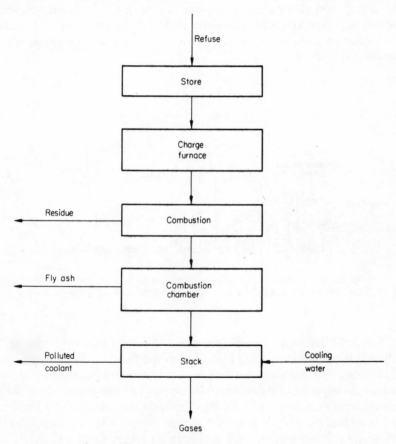

Figure 2.5:   Flow diagram for incineration

volume reduction which is of order 90 per cent of the original refuse volume and 60 per cent weight reduction is not uncommon.

Incineration can be recuperative, which means that heat recovery is practised, or non-recuperative, which is the more common in the UK. Many recuperative schemes are operated in areas where fuel is expensive (for example in Hong Kong). The basic elements of a non-recuperative incineration scheme are given in figure 2.5.

Incineration takes place in three stages, namely evaporation, distillation and combustion. Evaporation takes place in the furnace from both radiant and convective heat exchange and as more heat is added to the refuse charge, volatile hydrocarbon gases are released with ignition taking place around 700 °C. On further heating (and supplying the requisite amount of oxygen) the fixed carbon is consumed and converted to CO_2. The inert non-combustible matter which remains is discharged to the ash hopper for disposal. A cross section of a three grate municipal incinerator plant showing the drying and combustion zones is given in figure 2.6; other additions include a rotary kiln after the third grate for reduction of any residual carbonaceous material and a grit arrestor in the cooled flue gas stream for emission control.

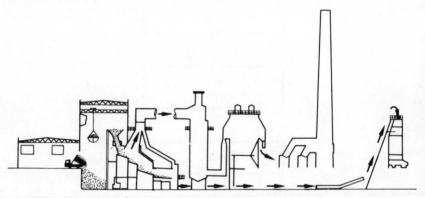

Figure 2.6:   Schematic Three-grate Municipal Incinerator with rotary kiln for complete burn-out. Courtesy of Clarke Chapman-John Thompson, Power Plant Division

It is seen that there can be four reject streams: incinerator residue, fly ash, polluted coolant and flue gases. It is noted that in preventing one form of pollution another must not be generated, thus strict control of all reject streams is required. This is especially so for the flue gases which should comply with the Clean Air Act 1956. The residue sterility is also important and is a measure of the plant performance. One test for the determination of fermentable matter in the ash is described in Appendix I on page 44.

The major problems with incineration and in fact any refuse disposal process are caused by refuse heterogenerity, two examples of this will suffice, namely plastics and aerosol canisters.

If the plastics are of the PVC variety (283,000 tonnes consumed in 1968 of which 150,000 tonnes was in disposable packaging) these problems can arise on incineration as hydrochloric acid is produced as shown in equation V below.

$$2[CH_2CHCl]_n + 5O_2 \longrightarrow 2HCl + 4CO_2 + 2H_2O$$

V

It is noted that similar reactions obtain for all halogenated plastics such as chlorinated polyethylene. The PVC disposal problem may become a vexing one for incinerator designers as the indications are that the PVC content of refuse may rise from its present 0.1–0.2 per cent by weight to 1–2 per cent by 1980.[6] Thorough flue gas cleaning will then be necessary probably with the use of scrubbers shown diagrammatically in figure 2.7 where the soluble HCl is contacted with water sprays to obtain efficient removal.

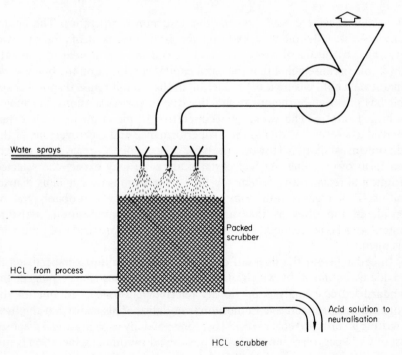

Figure 2.7:   Scrubber for HCl and grit removal

The incineration of aerosol canisters presents a problem to the incinerator designer also, as the propellant is commonly trichlorofluoromethane (CCl_3F) which decomposes on incineration to produce fluorides which have been blamed for seriously corroded incinerators in the US. Thus the artefacts of the throw-away age generate problems on disposal.

The state of the art for incineration can best be described by analysing the well-documented Edmonton incineration plant completed in 1971 for the GLC. This plant represents the latest in modern incineration practice and the claimed weight reduction is given in table 2.3 from a paper by Pepe & Turner.[7]

Table 2.3 *Combustion Performance of Edmonton Incinerator*

Crude refuse input	1,333 tons per day
Output: clinker	259 tons per day
metals	147 tons per day
fly ash	75 tons per day
Total residuals	481 tons per day
Percentage weight reduction	64

Two schemes were costed, recuperative and non-recuperative. The actual Edmonton installation was made recuperative, that is steam raising was performed by means of waste heat boilers and the steam used to generate power, the premise being that the extra capital employed on the boilers and generators would ensure a net reduction in the overall refuse disposal costs. This has not been borne out in practice due to cost escalations and major teething troubles. The cause of recuperative incineration in the UK has received a severe blow due to the malperformance and cost overruns of the Edmonton installation. However the basic concept of heat recovery is correct as a form of recycling. As fuel costs escalate we may expect the selected adoption of recuperative schemes in the UK for various uses such as district heating. Power generation from refuse incineration will probably not be considered too often as the variations in refuse composition, moisture content and boiler reliability all militate against the provision of power by this means.

In order to put the disposal costs of incineration into perspective and provide a cost norm by which other disposal processes can be judged, the standardised cost analysis previously referred to was carried out on the proposed non-recuperative Edmonton installation. This showed at the time of writing an installed cost of £5,730 per tonne of daily capacity and a disposal cost of £3.34 per ton of input refuse. Thus the adoption of incineration is not a negligible step and is only undertaken when controlled tipping space is at a premium.

Cellulose hydrolysis

Cellulose hydrolysis applied to refuse disposal is a new process proposed by the writer[5] with materials recycling in mind. So far as is known, this process has not been previously applied or proposed for refuse disposal although there are recorded applications of it to cellulosic materials such as cotton hulls and hardwoods.

The hydrolysis process is based on two consecutive first order chemical reactions.

$$A \xrightarrow{k_1} R \xrightarrow{k_2} S$$

Where: A ---- Cellulose
R ---- Fermentable sugars
S ---- Decomposed (non-fermentable) sugars
k_1 ---- Reaction rate constant, cellulose to fermentable sugars
k_2 ---- Reaction rate constant, fermentable to decomposed sugars

VI

It is noted that k_1 and k_2 are highly temperature dependent. As equation VI shows, the fermentable sugars are subject to decomposition on continued exposure to the hot dilute acid, the reactions being rate controlled. Thus for any given hydrolysis conditions, that is acid concentration and temperature, there is an optimum reaction time for maximum fermentable sugar yield after which the temperature must be sharply reduced to quench the reaction and stabilise the yield. It can be shown that for high sugar yields and low costs the reaction should be carried out at high temperature and low acid concentration and hydrolysis conditions of 0.4 per cent H_2SO_4 concentration and 230 °C were found to be the approximate upper limits for a controllable reaction. The optimum residence time for maximum yield under these conditions is 1.2 minutes with a 55 per cent conversion to fermentable sugars. The concentration versus time profile for a continuous reactor using the above conditions is shown in figure 2.8. It is noted that 23 per cent of the original cellulose is unhydrolysed and if efficient separation can be obtained from the sugar liquor this fraction may be recycled through the reactor.

Once the fermentable sugars are obtained a variety of processes can be carried out, the chief one being fermentation which, depending on the yeast used, can yield an aqueous solution of ethyl alcohol (butanol, citric acid, fodder, yeast are other possible fermentation products). The aqueous ethanol solution is subsequently rectified and an azotropic mixture of 95 per cent ethanol and 5 per cent water obtained. The fermentation reaction to ethanol alcohol has already been given in equation II.

Figure 2.9 gives a rudimentary flow diagram for municipal refuse hydrolysis which illustrates the principal steps employed. Briefly, the refuse is

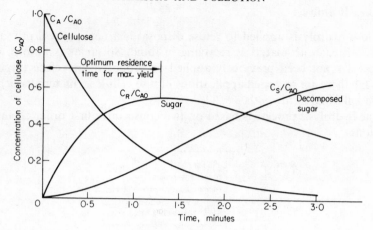

Figure 2.8: Concentration-time profiles for the continuous hydrolysis of
cellulose at 230° and 0.4 per cent H_2SO_4

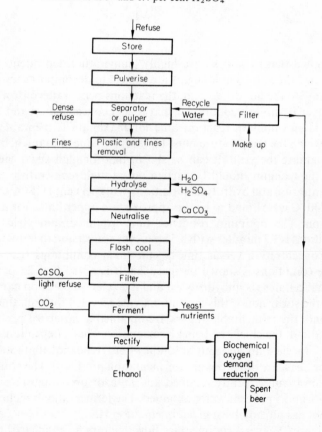

Figure 2.9:   Flow diagram: ethanol from refuse

pulverised and discharged into a flotation separator or a special pulper to allow segregation of the refuse into a dense and light fraction. The pulped fraction, mainly cellulosic materials, is then fed to a fines and plastic removal section and thence to a reactor for acid hydrolysis at 230 °C and 0.4 per cent H_2SO_4 with an optimum residence time of 1.2 minutes to obtain maximum conversion to fermentable sugars, flash cooling using the process feed water as coolant, neutralisation with $CaCO_3$ and filtering flow. Fermentation is then carried out for roughly 20 hours at 40 °C and the resulting 1.7 per cent aqueous ethyl-alcohol solution is distilled or rectified to yield 95 per cent ethyl-alcohol and 5 per cent water. A waste liquid stream is also discharged which requires treatment for biochemical oxygen demand reduction.

This then is the proposed process. It is conceptually simple but much has to be done before the commercial stage is reached. Preliminary work done under the auspices of a US Public Health Service Grant[8] and reported by Fagan et al. has shown that the kinetics of the acid hydrolysis of figure 2.6 are substantially correct. Fermentable sugar yields of around 45 per cent were obtained at hydrolysis conditions similar to those used in the analysis. Further work is in progress.

Standardised costing procedures[5] showed that a 250 tonne per day plant with a 40 per cent paper content would show a profit of £0.383 per tonne of input refuse and on a 60 per cent paper content a profit of £2.72 per tonne. The capital cost was estimated at £5,870 per tonne of daily installed capacity. The assumptions made are that the predicted ethanol yields are

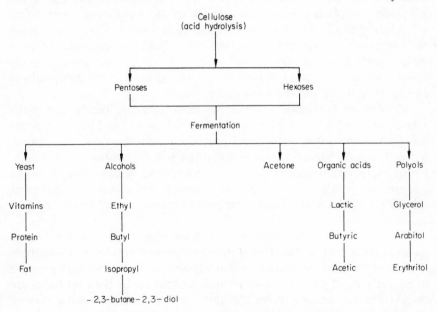

Figure 2.10: Fermentation products of cellulose hydrolysis

obtained in practice and that the ethanol market price structure remains stable.

It is to be noted that this process is an attempt to recycle a refuse constitutent in a useful form and is indicative of the direction that refuse processing will take in the future when waste is no longer material for which a use cannot be found. With the current escalation in crude oil prices there is now strong interest in ethanol production by the fermentation route and little prospect of market restrictions. Figure 2.10 shows other possible fermentation products.

One aspect which has not been actively considered so far is the production of artifical protein from cellulose in refuse. The process requirements are identical as the starting point for the production of the protein is a supply of fermentable sugars. Again this promising concept may receive more attention when the full impact of a resource hungry world is felt.

Obviously other cellulose recycle processes are possible and one such process is pyrolysis of refuse to produce combustible gases and, hopefully, oils as described in the next section.

Refuse pyrolysis and oil production

In the processing of refuse or for that matter in the solution of any waste disposal problem a new form of pollution must not be generated while effecting disposal, that is there must be no conversion of pollution from one form into another; for example, incineration of waste can give rise to air pollution, hydrolysis of refuse can give rise to water pollution—both forms of pollution can be tackled provided money is spent in the process. Both processes (incineration and hydrolysis) have substantial quantities of solid waste which also require disposal after processing. One process which appears to have a strong potential for producing both usable end products and effecting a safe and clean disposal is pyrolysis.

Pyrolysis or destructive distillation of refuse is the indirect heating of the refuse in a retort at elevated temperatures in the range 250–1,000 °C depending on the processing techniques employed and the products desired. Virtually any carbonaceous material that can be volatilised can be pyrolised. From the refuse disposal viewpoint, the wastes can be converted to usable or more manageable solid, liquid and gaseous forms with a lower potential for contributing to land, air and water pollution. (See sections on incineration and hydrolysis.)

To date pyrolysis of refuse is at the pilot plant stage and two distinct patterns are being followed, that of pyrolysis 'proper' to recover combustible gases, and a related technique where the refuse is heated under high pressure in the presence of CO and water to produce heavy oils. Both processes are discussed as the writer believes that this form of refuse processing may be of considerable importance in the future.

The refuse is heated in a retort to solid, gaseous and liquid fractions. Substantial work is being conducted in the us and the analysis of refuse used in the tests of Sanner et al.[10] are given in table 2.4 below.

Table 2.4 *Average analyses of US refuse used in pyrolysis (destructive distillation) tests*

	Raw municipal refuse	
	As received	Dry
Proximate, percent:		
Moisture	43.3	—
Volatile matter	43.0	76.3
Fixed carbon	6.7	11.7
Ash	7.0	12.0
Total	100.0	100.0
Ultimate, percent:		
Hydrogen	8.2	6.0
Carbon	27.2	47.6
Nitrogen	.7	1.2
Oxygen	56.8	32.9
Sulphur	.1	.3
Ash	7.0	12.0
Total	100.0	100.0
Btu per pound of refuse	4,827	8,546
kJ/kg of refuse	11,200	19,800
Available Btu per ton of refuse, millions	9.654	17.092
Available kJ/kg refuse	9,900	17,500

Analysis of refuse used in Pyrolysis tests. us Bureau of Mines Report—7428.

Pyrolysis was carried out at two constant temperatures of 750 °C and 900 °C. The products and their yields are given in table 2.5.

Table 2.5 *Yield of Products from Pyrolysis of 1 tonne of refuse*

Char	154–230 lb	70–104 kg
Tar and pitch	0.5–5 us gall	1.9–19 l
Light oil	1.5–2 us gall	5.65–7.5 l
Ammonium sulphate	18–25 lb	8.15–11.3 kg
Liquor	80–133 us gall	302–533 l
Gas (15 °C, 1 × 105 Pa)	11,000–17,000 cu ft	314–486 m^3

The char obtained had characteristics which varied widely depending on the refuse. Municipal refuse with a high plastics content yielded a char with 6.7 per cent fixed carbon whereas industrial refuse produced a char with roughly

12 per cent fixed carbon. The sulphur content was less than 0.2 per cent in all tests.

The gas content is most interesting. The major constituents are hydrogen, methane, carbon dioxide, carbon monoxide and ethylene, in that order. An analysis is given in table 2.6.

Table 2.6 *Gas Analysis from Pyrolysis of Municipal Refuse* (*Analysis vol per cent*)

Pyrolysis temp °C	750	900
Hydrogen	31	52
Carbon monoxide	16	18
Methane	23	13
Ethylene	11	5
Carbon dioxide	19	12
Btu/cubic foot gas	563	447
kJ/m^3	45,500	36,300
Million Btu/ton refuse pyrolised	5.421	7.930
kJ/kg/refuse pyrolised	5,600	8,200

The gas analysis in table 2.6 is in line with the pyrolysis work reported by the Karl Kroyer Coy[11] in Denmark.

What is interesting is that at the higher pyrolysis temperature of 900 °C a significantly higher yield of combustible gas is obtained compared with the lower temperature of 750 °C. Tests on various grades of refuse have confirmed that these higher yields are always obtained at the higher temperature. Now the energy requirements for the pyrolysis of 1 tonne refuse is roughly 2.1×10^6 kJ (2 million Btu) and as approximately 8.4×10^6 kJ (8 million Btu) are available per tonne at 900 °C (based on US refuse compositions) then the process is self-sustaining in energy requirements and has a surplus which can be used to augment town gas supplies. This has been done in the town of Kolding in the 'Destrugas' pilot plant operation.[11] Figure 2.11 shows a schematic diagram of the Kolding plant. Gas production by refuse pyrolysis is now available commercially and it will be interesting to see the extent to which this method is adopted.

Domestic refuse, cellulosic wastes (and even sewage sludge) have been converted to heavy oils by heating under pressure with carbon monoxide and steam. The techniques and results are reported by Appell et al.[12,13]

Basically the process requires temperatures of 250°–400 °C and pressures of 1×10^7–3×10^7 Pa respectively. The yield of oil based on dry organic matter is roughly 40 per cent irrespective of the processing condition in the 250–400 °C range, which is the equivalent of 238 l (2 barrels of oil) per tonne of US refuse. Given the usual lower paper content of UK refuse, the UK yield would be 178 l (circa 1.5 barrels) per tonne of refuse. However, the oil characteristics vary greatly. The oil obtained at 400 °C flows slowly at room

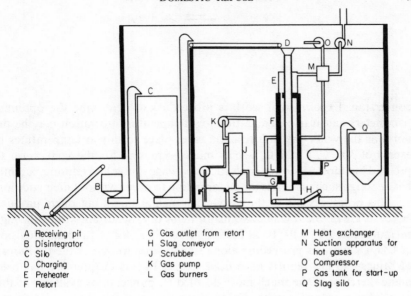

A Receiving pit   G Gas outlet from retort   M Heat exchanger
B Disintegrator   H Slag conveyor   N Suction apparatus for
C Silo   J Scrubber   hot gases
D Charging   K Gas pump   O Compressor
E Preheater   L Gas burners   P Gas tank for start-up
F Retort   Q Slag silo

Figure 2.11: Destrugas Process (ref. 11)

temperature whereas the 250 °C product is a soft solid and must be warmed before flow takes place.

As with the pyrolysis process, oil production can effect a virtually pollution free disposal and recover a commercial product, in this case, oil.

The results of tests carried out on the effectiveness of carbon monoxide and steam in the conversion of the content of refuse to oil at 350 °C are shown in table 2.7. The results obtained with hydrogen are shown for comparison.

Table 2.7 *Composition of benzene soluble oils from refuse* (2 *hours at* 350 °C, 1×10^7 Pa *initial pressure*)

Gas	$CO + H_2O$	H_2
Refuse conversion per cent (weight)	40	5.0
Carbon per cent	83.3	76.3
Hydrogen per cent	7.8	8.2
Oxygen per cent	8.9	15.5

It is significant that the addition of carbon monoxide and water give high conversions whereas the addition of hydrogen does not. The exact function of the CO is not known, it is possible that it reacts with water and alkaline salts present in the refuse to form some intermediate compound which transfers hydrogen, probably as hydride ion to the substrate thus leading to oil formation. On the other hand, it could remove oxygen from the cellulose by formation of carbon dioxide. It transpires that the water-gas shift reaction shown in VII must be minimised as it consumes the CO and inhibits the

$$CO + H_2O \rightleftharpoons CO_2 + H_2$$

VII

conversion. Experimental work is proceeding to determine the optimum parameters and it is probable that low temperature operation may be the norm as the water gas shift reaction takes place readily at temperatures in excess of 300 °C. Catalytic agents must be present for the conversion to proceed but most waste materials such as sewage sludge and refuse apparently have enough such materials for oil formation to occur without their addition.

This process is still in the early experimental stages but offers promise. No costs are obtainable but two major items of expenditure will be the high pressure reactor (1×10^7 Pa at 250 °C, 3×10^7 Pa at 400 °C) and the provision of CO. UK work is proceeding along similar lines to that of the US Bureau of Mines—so far, reports have been sparse. In this chapter use has been made therefore of the much more detailed US publications available in this area. That pyrolysis is in its infancy cannot be emphasised too strongly but on an energy costing basis it would appear that, based on UK refuse compositions, half the available energy can be sold.

Trends

Refuse composition will tend to more paper and plastics as the throw-away age accelerates. Due to the unsuitability of much of the paper in domestic refuse for repulping on a commercial basis, recycling effort will be directed at the conversion of the organic content into useful products such as protein, ethanol, gas or oil. The reasons for this are not hard to seek. Figure 2.12 from *Resources and Man*[14] shows two well-founded predictions for the complete cycles of world crude oil production based on two estimates of the ultimate quantity Q_∞ which it is possible to obtain. The world energy outlook based on oil fuels by the turn of the century is bleak indeed and increasing emphasis will certainly be placed on the processing of refuse as a replenishable energy resource. Refuse will soon be too valuable to share with holes in the ground.

Appendix I

Dusseldorf method of incinerator residue analysis for the determination of unburnt carbon and fermentable matter after Harvey[15]

The sample is ground and dried for one hour at 105 °C to remove the moisture content so that the subsequent analysis can be conducted using the dry substance.

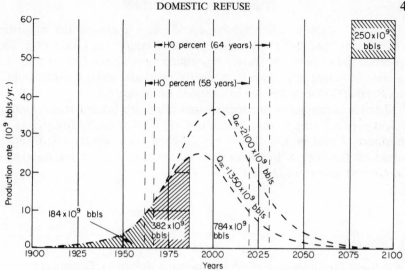

Figure 2.12: Complete cycles of world crude-oil production for two values of Q_0 (ref. 14) (US billion barrels)

In order to hydrolyse all fermentable substances, 200 g dry substance is mixed with 500 ml of 0.3 per cent NaOH in a 1,000 cc beaker. The whole is refluxed for three hours and subsequently cooled to ambient temperature. If necessary, make up to the initial level with distilled water and mix thoroughly.

The solution is drawn off by filter suction and 250 ml (= 100 g of weighed sample) is retained.

20–30 ml concentrated H_2SO_4 is added to 250 ml of this filtrate in a round-bottom flask and refluxed until the solution has settled. This results in all carbonate residues being decomposed and the CO_2 being boiled off (approximately $1\frac{1}{2}$ hours). In a closed CO_2 apparatus, 50 ml of 0.1 N $KMnO_4$ solution is added to the 250 ml test solution via separating funnel. The solution must be coloured violet and if necessary the quantity of permanganate is increased. The reaction with $KMnO_4$ takes place as shown below (VIII), sodium oxalate was included in the equation to demonstrate the fermentable substance.

$$5\ Na_2C_2O_4 + 2\ KMnO_4 + 8\ H_2SO_4 = K_2SO_4 + 5\ Na_2SO_4 + 2\ MnSO_4 + 10\ CO_2 + 8H_2O.$$

VIII

The liquid is heated to boiling while air is passing through. The air passing through the liquid is first purified by passing it through a 50 per cent solution of KOH and a sodium-asbestos filter. The gas stream containing the

CO_2 is sucked through two sulphuric acid filters to absorb any impurities. A $CaCl_2$ filter and a P_2O_5 filter absorb the moisture. The total CO_2 is then absorbed in two sodium-asbestos absorption vessels arranged in series. The increase in weight of the two vessels is recorded and related to the weighed sample, then converted to the fermentable substance.

The maximum allowable fermentable substance is fixed at less than 0.3 weight per cent, that is, the amount of CO_2 formed finally from 100 g dry substance should be less than 0.575 g. When the content of fermentable substance falls below this figure no cultures are formed on Agar-Agar. Conversion factor: weight of $CO_2 \times 0.52$.

References

1. City of Birmingham (1970). Private communication (December)
2. A. Porteous (1969). The recovery of industrial ethanol from paper in waste. *Chemistry and Industry*, 6th December, pp 1763–1770
3. K. C. Dean *et al.* (1971). *Preliminary Separation of Metals and Non Metals from Urban Refuse.* USBM Solid Waste Research Programme Tech. Prog. Report 34 (June)
4. K. R. Gray and A. J. Biddleston (1972). Composting-Process Parameters. I.Ch.E. Symp. *Disposal of Solid Wastes* (27th September)
5. A. Porteous (1971). A new look at solid waste disposal. *Institute of Public Cleansing*, **LXI** (4), April, pp 152–169
6. J. J. P. Staudinger. *Disposal of Plastics, Waste and Litter.* Society of the Chemical Industry, S.C.I. Monograph No. 35.
7. P. D. Pepe and G. M. Turner (1968–9). Design of the first large UK power producing refuse disposal plant. *Proc. Institution of Mechanical Engineers*, **183**, part 1 (24)
8. US Public Health Service Grant No. U1 00597–01 (1968). *Kinetics of the Porteous Hydrolysis Process—Hydrolysis of Paper.* Awarded to the Thayer School of Engineering, Dartmouth College, Hanover, N.H., USA
9. R. P. Fagan *et al.* (1970). *Kinetics of the Acid Hydrolysis of Cellulose found in Paper Refuse.* Northeast Regional Anti-Pollution Conference, Kingston, R.I., USA (September)
10. W. S. Sanner *et al.* (1970). *Conversion of Municipal and Industrial Refuse into Useful Materials by Pyrolysis.* USBM Solid Waste Prog. Report 7428 (August)
11. Karl Kroyer Coy (1970). Pyrolysis of Waste. *Stads – og havneingenioren* (5)
12. H. R. Appell *et al.* (1970). *Conversion of Urban Refuse to Oil.* USBM Solid Waste Programme Tech. Prog. Report 25 (May)
13. H. R. Appell *et al.* (1971). *Converting Organic Wastes to Oil.* USBM Report RI 7560

14. *Resources and Man* (1969). National Academy of Sciences and National Research Council Report. W. H. Freeman, San Francisco

15. K. Harvey. Incineration Practical Aspects (1972). I.Ch.E. Symp. *Disposal of Solid Wastes* (27th September)

CHAPTER 3

PESTICIDES AND POLLUTION

F. R. BENN and C. A. McAULIFFE
Department of Chemistry, University of Manchester Institute of Science and Technology, Manchester M60 1QD

The other subjects of this book are pollutants which stem primarily from the highly developed industrial societies, but pesticides, which are used in large quantities throughout the world, open up the possibility of global pollution. They have certainly made an enormous impact since the introduction of the modern synthetic organic pesticides. DDT is now the most notorious, although its value in reducing disease—spectacular in the arrest of the Naples typhus epidemic during World War II, but more far reaching in the reduction of mosquito and consequent erradication of malaria from large tracts of the world—is priceless. Of course, the increased life span and improved quality of life which it has given to so many people must also be viewed in the light of the already disastrous increase in world population, for it is in the very areas in which pesticides make their greatest impact that population growth is greatest and food supplies most scarce.

Since pesticides are so closely linked with food supplies and population growth, it is worthwhile examining the effect of the latter in detail.

We can express the rate of growth of population thus:

$$N_t = N_o e^{rt}$$

where N_o = population at time O,
N_t = population at time t,
r = growth rate (the difference at any instant between the number of people being born and the number dying)
and t = time in years.

Thus, for a doubling of the population, $N_t/N_o = 2$

$$2 = e^{rt}.$$

Taking the natural logarithm of each side

$$\ln 2 = rt$$

$$\frac{\ln 2}{r} = t \qquad \text{or} \qquad \frac{0.6931}{r} = t$$

Then, for a growth rate of 2 per cent, r = 0.02

$$t = \frac{0.6931}{0.02} = 34.65 \text{ years}$$

Table 3.1 *Population Doubling Time*

Growth rate (% per year)	Doubling time (years)
0.1	700
0.5	140
1.0	70
2.0	35
4.0	18
5.0	14
7.0	10
10.0	7

Table 3.2 *Average Annual Growth Rate of Population 1961–68*

	(% per year)
India	2.5
USSR	1.3
USA	1.4
Pakistan	2.6
Indonesia	2.4
Japan	1.0
Brazil	3.0
Nigeria	2.4
W. Germany	1.0
UK	0.6

Source: World Bank Atlas (Washington DC: International Bank for Reconstruction and Development, 1970)

The population growth is determined not only by the birth rate, a popular misconception, but by the death rate as well, and it is of course the difference between these two which decides the population growth. While man has had little success in controlling birth rate in the countries where it matters most, his impact on the death rate has been astounding.

The results in table 3.1 and 3.2 show that in the UK, where there has been so much fuss about over-population, the problem is insignificant in comparison with the countries of the eastern hemisphere. If we cite the case of Ceylon, a country which in 1945 had a death rate of twenty-two per thousand, the use of DDT for control of malaria-carrying mosquitos and the introduction of other health improvement measures reduced the mortality to less than half in a decade.

Unless we can control birth rate and at the same time increase the quantity of food for the inevitable extra mouths, then the benefit of saving lives will become the tragedy of famine.

Pesticides make an obvious contribution to improving food production although it is not generally realised how great is their potential and how dependent we already are on them for the maintenance of our present food supplies. If we accept that the world population will double in the next thirty to forty years and yet at the present time only one third of mankind has sufficient food, it can be seen that the world as a whole is already in a bad way. On the basis of UN forecasts, it is predicted that food supplies must be trebled by the end of the century in order to provide a reasonable level of nutrition throughout the world.

There is plenty of room for improvement in agricultural food production. As much as one third of the potential crop may be lost up to harvest time due to the effects of insects, weeds, plant diseases, etc., and similar losses may occur during storage. Pesticides alone cannot eliminate these problems, but because of their great success so far they are bound to be included in any improvement programme which is to be feasible in the foreseeable future, together with other major items such as the use of new crop strains and more sophisticated management of agriculture. Accurate estimates of crop losses and the remedial effectiveness of pesticides are difficult to obtain[1] but it is probable that the majority of the loss can be saved by suitable crop protection.

Accepting the need for pesticides in the spheres of public health and agriculture, it is also evident that they have at times been over-used and

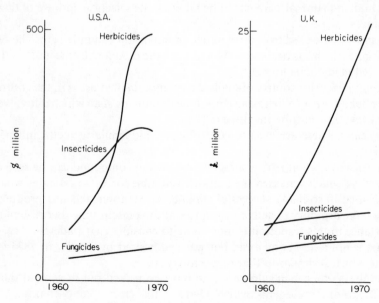

Figure 3.1:   Trends in pesticide sales[2,3]

misused to an extent that has given rise to public concern because some at least have led to widespread and long-term contamination of the environment. The problems are relatively new since the highly active synthetic organic pesticides, the very ones producing most of the current environmental problems, have been developed during the last thirty years and their use has grown rapidly. Some idea of their importance is shown in the sales graph (figure 3.1) for the major groups of insecticides used in the USA and the UK.

Pesticides may be divided into more or less well-defined groups, the three major ones being insecticides, herbicides and fungicides. However, there are many other groups which are quantitatively less important but which still make a vital contribution to the total effort. Some of these are also listed below.

Insecticides. This is a major group containing many different classes of compounds and giving rise to most of the environmental problems which have been met.

Herbicides are weed killers. Once again this is a major group comparable in economic importance with the insecticides.

Fungicides. Control of fungal infections has been dominated in the past by the use of sulphur and copper compounds and it is only recently that any real impact has been made by synthetic organic compounds. The economic importance of this group is less than that of the insecticides and herbicides.

Rodenticides, etc. These are substances used against undesirable mammals such as rats. Because of the similarity of their metabolic processes with that of the human mammal, care has to be taken in the selection and use of these substances.

Acaricides are used to control mites, an important example being the red spider mite which has become prevalent since the organochlorine insecticides destroyed many of its predators.

Molluscicides for control of snails. An example of their use is in the control of liver fluke disease (bilharziasis) by killing the aquatic snail which is involved in the cycle transmitting the disease.

Algicides. These are used to control algae and similar growths in water systems.

The discovery that DDT is a highly insecticidal compound can be seen as the catalyst which activated the fantastic scramble for new pesticides which has resulted in so many successful commercial ventures and has produced such a large armoury of active substances at the present time. Because of its importance in this respect, it is interesting to consider DDT in detail.

DDT is not a new compound but was synthesised by Zeidler in 1874 via a route which is similar to that in use today (I).

Catalysts other than sulphuric acid have been used but in general these are all protonating agents which increase the electrophilic character of chloral. The mechanism of the reaction is represented below (II).

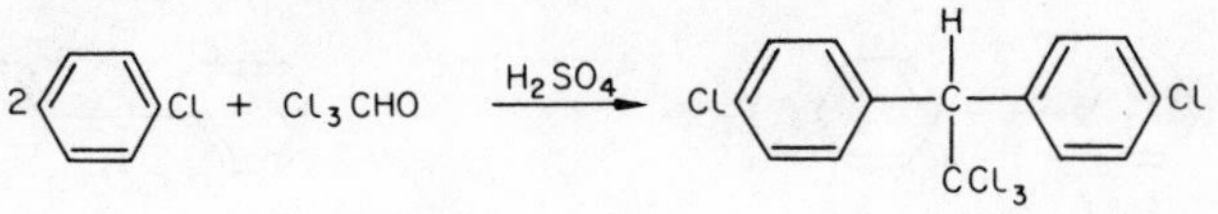

DDT
1, 1 – bis (4 – chlorophenyl) – 2, 2, 2–
trichloroethane

I

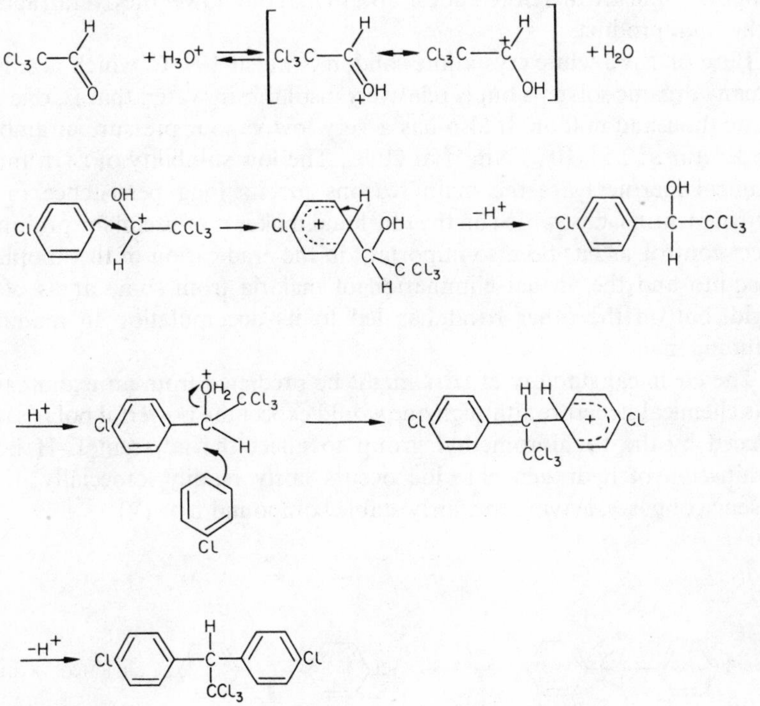

II

Reaction takes place readily at about 60 °C so that the production of DDT requires only simple manufacturing plant. As might be predicted from the known behaviour of chlorobenzene in electrophilic substitution reactions, substitution although occurring mainly at the 4 position, also takes place at the 2 position so that smaller quantities of the other expected isomers are present (III and IV).

C

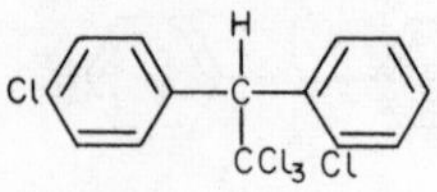

2, 4′– isomer   2, 2′– isomer

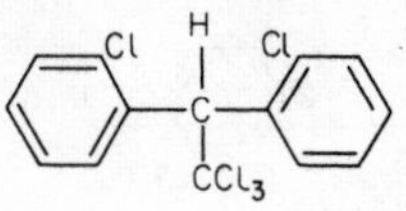

III   IV

Both of these compounds have lower insecticidal activities than the parent compound and are therefore undesirable in that they lower the overall activity of the total product.

Pure DDT is a white crystalline solid, melting at 109 °C, which is soluble in many organic solvents but is relatively insoluble in water, that is, one part in one thousand million. It also has a very low vapour pressure at ambient temperatures: $2.5 \times 10^{-4}\,Nm^{-2}$ at 20 °C. The low solubility of DDT and its chemical inertness are the main reasons for its long persistence in the environment, a factor which on the one hand makes it so useful for prolonged insect control, as has been so important in the eradication of the anopheles mosquito and the virtual elimination of malaria from some areas of the world, but on the other hand has led to its accumulation in mammals including man.

The chemical stability of DDT might be predicted from an examination of its chemical structure although one would expect the powerful polarisation induced by the trichloromethyl group to affect the adjacent C-H bond. Elimination of hydrogen chloride occurs fairly readily, especially in the presence of bases, leaving the fairly stable compound DDE (V).

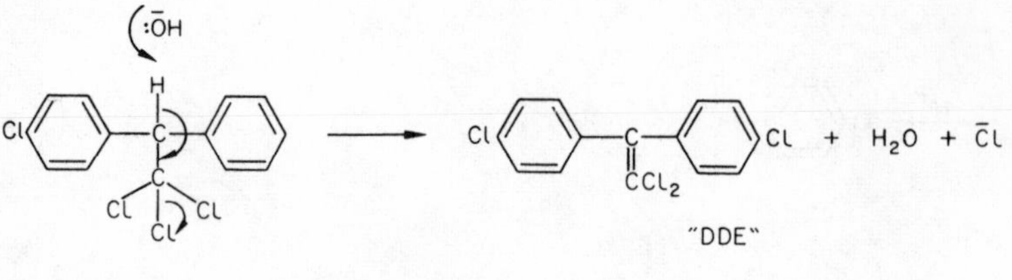

"DDE"

V

This dehydrochlorination also occurs in the presence of metal salts and in soil so that the stable residues which are found after spraying DDT normally contain DDE. Certain resistant strains of insects have an enzyme system which will carry out the same function and so render the insecticide inactive, for DDE is not toxic towards insects.

There are many analogues of DDT which have similar insecticidal activity and some of them have found commercial application in special cases. Two which are of particular interest in connection with resistant species are 'Deutero DDT' (VI) and 'Orthochloro DDT' (VII).

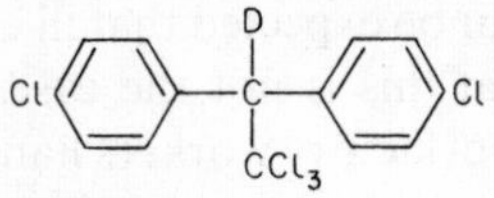

"Deutero DDT"   "Orthochloro DDT"

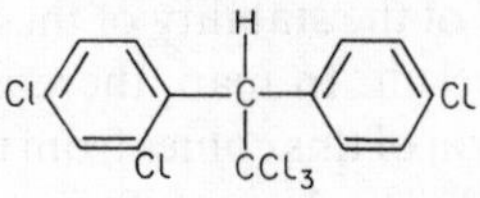

VI   VII

Both these compounds have shown activity against DDT-resistant insects, and it is interesting to speculate how selective the detoxifying enzyme system must be to be affected by such small changes as are shown in these molecules compared with the parent DDT. In fact, although the mode of action of DDT and related insecticides is not fully understood the structural requirements do not allow much change from the basic shape of DDT.

Chlorine in the 4 position may be replaced by fluorine, methyl and methoxyl without any major change in insecticidal activity. The methoxyl analogue methoxychlor has found use as an insecticide on animal forage because it is not accumulated in fatty tissues and will not therefore contaminate cows' milk (VIII).

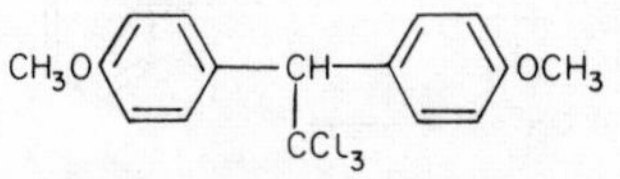

Methoxychlor

VIII

DDT is a broad spectrum insecticide; that is, it is effective against a wide variety of insects not all of which are harmful, so that like many other conventional synthetic organic insecticides, it destroys many pests, but affects some useful insects, hence the search in recent years for more specific remedies. It is an extremely active compound having an LD_{50} in the range 10^1 to 10^2 $\mu g/g$. That is, the dose necessary to kill 50 per cent of a treated population of susceptible insects is between 10 and 100 micrograms per gram of insect body weight.

Although DDT is so toxic to insects, producing extreme activity, exhaustion, and finally death, virtually no ill effects have been noted in man, which is surprising considering that the amount which has been used is equivalent to about one pound for every person on earth.

DDT and DDE concentrate in the body's fatty tissue and in the liver, and because of the stability of these substances it might be expected that ill effects would result. In man, there is some evidence that this is not the case.[4] An indication of this comes from the fact that a group of factory workers handling DDT and absorbing sufficient to maintain a level of thirty to sixty times that present in the population at large showed no untoward effects during a ten year period. On the basis that drug response is proportional to the product of dosage and time, it has been predicted that a man would have to live about 300 years before any symptoms showed at present body levels of DDT. On the other hand, like so many substances in common use today which have been accepted as non-toxic for many years, it has been suggested that DDT might be carcinogenic. Although the evidence is very meagre, it seems to have achieved some importance in the recent discussions on the use of DDT.[5]

The level of DDT in man varies quite widely from country to country. In the UK, it is about 1 ppm (in fat), in the USA about 2 ppm, but in India the concentration has been higher than 10 ppm, reflecting the differing use of pesticides in these countries.

Its presence in body tissue is not static although it may persist for a decade or more. However, it is gradually excreted either as DDT or one of its metabolic products such as DDE, or the product of hydrolysis, DDA. The latter may be obtained in the laboratory as its sodium salt by heating DDT for a long time with sodium hydroxide solution (IX).

IX

DDT is relatively cheap, costing approximately one dollar per pound of active ingredient in the formulated product, and this coupled with its lack of acute mammalian toxicity has commended it to the developing countries because of their abundance of low cost but unsophisticated labour which is incapable of handling the more toxic materials safely. In these cases the relatively remote possibility of long term toxic effects is insignificant in comparison with the great immediate benefits of using this insecticide.[6]

Although there is little evidence to suggest that DDT is harmful to man, the concentration of DDT and other chlorinated hydrocarbons in food chains must, at least, give rise to disquiet.

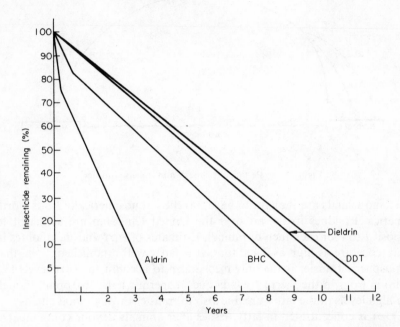

Figure 3.2: Trends in pesticide retentions

Figure 3.2 gives an indication of the average survival of a number of chlorinated hydrocarbons in soil. Nash and Woolson[7] have estimated the persistence of chlorinated hydrocarbons in soil, and the probable upper limit to persistence is shown in table 3.3.

Table 3.3

Pesticide	Time	% remaining
Aldrin	14 years later	40
Endrin	,,	41
BHC	,,	10
Aldrin	15 years later	28
Dieldrin	,,	31
DDT	17 years later	39

The transfer of pesticides over long distances is generally via atmospheric movement. Pesticides can enter the air via vaporisation or codistillation with water, and up to 50 per cent of pesticide residues have been claimed to enter the atmosphere in this manner. A remarkable study by Riseborough et al.[8]

has shown that pesticide residues from Morocco have been transferred atmospherically to Barbados (figure 3.3). Pesticide residues detected in the

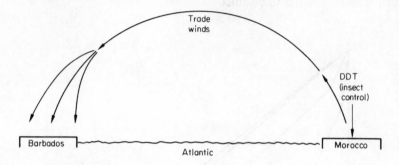

Figure 3.3:   Pesticide residues—atmospheric transfer

Shetland Islands are thought to have travelled atmospherically from North America. Residues in the air over the United Kingdom are sufficient to deposit 1 ton for each inch of rainfall. Estimates of DDT and metabolites in Antarctica are as high as 2,400 tons. It is fairly well established, then, that atmospheric transfer is the only mechanism to account for DDT levels (3.2 ppm) in frogs in the Sierra Nevada and in sperm whales (6.0 ppm), though the high levels in the latter are due also to transfer through food chains.

DDT is concentrated in fatty tissues after animals devour contaminated flora. Losses of DDT along food chains are virtually unknown; invariably it is concentrated. Oysters, which live by filtering water, can concentrate up to 70,000 times the amount of DDT in the water in which they exist. Thus, the next creature along the food chain which consumes the oyster will obtain this increase and, presumably, concentrate it further. This provides a some-what startling example, but illustrates the general principle.

DDT is the first member of a group of compounds known as the organo-chlorine insecticides which have the common property that they are relatively unreactive chemically speaking and are therefore stable and persistent. These compounds can and do concentrate in certain forms of wild life particularly the creatures which are at the head of carnivorous food chains. In these cases there is strong evidence that toxicity does occur and the long-term effects on bird life in particular and wild life in general are such as to make the risks both agriculturally and aesthetically unacceptable.

There has been a pronounced tendency in recent years to judge the whole group of compounds by the combined shortcomings of individual com-pounds and to ignore their separate identities, with the inevitable con-sequence of a considerable lobby demanding their total withdrawal. Such an action is unlikely to seriously affect the wealthier nations of the world since many alternative pesticides have become available in recent years, although these do not in general match the low cost of DDT. The poorer nations with severe public health problems are likely to be concerned more with their immediate problems than with long term environmental effects, and in these cases the relatively safe and cheap organochlorine compounds will continue in use.

Soon after the introduction of DDT, other compounds containing chlorine were shown to be insecticidal. Gamma benzene hexachloride (γ-BHC) from France and the UK, and the cyclodienes, aldrin and dieldrin, from the USA, are of note although many other chlorine-containing compounds became important because of their relatively low cost and ease of manufacture.

γ-BHC is a very simple compound resulting from the addition of chlorine to benzene in the presence of peroxides or photochemically by irradiation with ultra-violet light (X).

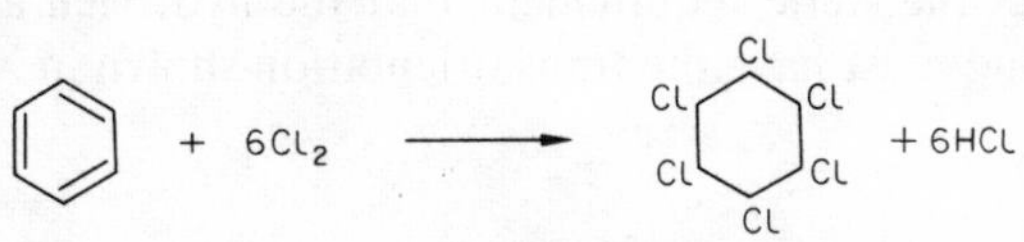

X

Unfortunately the stereochemistry allows the formation of eight isomers, only one—the so-called γ-isomer—having significant insecticidal activity, and this is present to the extent of 10–18 per cent in the reaction product. The active compound has three adjacent C-Cl bonds axial to the ring, the remainder being equatorial (XI).

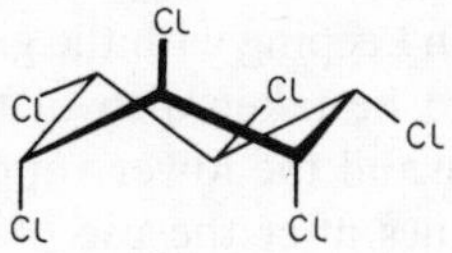

XI

In spite of this low yield of the required material the cost is not excessive, partly because of the low raw material cost and also because the by-products can be converted to useful materials such as 2,4,5-trichloro-phenol, the basis for the herbicide 2,4,5-T.

γ-BHC has found particular use in agriculture to eradicate soil pests, but unless it is pure root crops become tainted with an unpleasant musty taste.

BHC is not as persistent as DDT although it has similar characteristics such as fat solubility, etc. Its degradation, whether in the soil, in plants, or in mammals, results in the formation of aromatic compounds probably via an initial elimination reaction shown in XII.

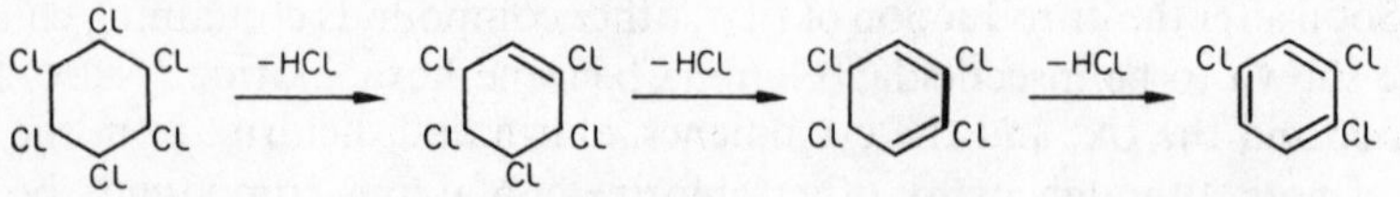

XII

Many other aromatic compounds are also found. The primary reaction is probably analogous to the commonly met base-initiated bimolecular elimination which has the steric requirement that the hydrogen attacked and the chlorine leaving must have the trans orientation shown in XIII.

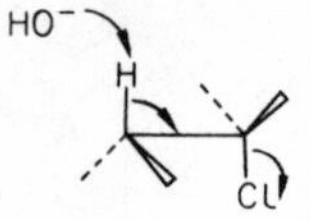

XIII

This is possible with all but one of the isomers of BHC. The β-isomer has all its chlorines in the equatorial configuration so that this trans relationship does not obtain. This is in keeping with the greater persistence of this isomer, although account should be taken also of the contribution to this of the lower solubility in water and the lower vapour pressure. As a consequence the environmental residues after the use of technical grade BHC contain a high proportion of the β-isomer.

The cyclodienes which form the other major part of the organochlorine insecticides are more complex in structure especially when one examines the possibility of stereoisomerism. In fact as with BHC this latter plays an important part in deciding levels of both insecticidal activity and mammalian toxicity. Like the previous compounds, they are stable and fat soluble and therefore pose serious problems of persistence and contamination of wild life. The two major compounds of the group, aldrin and dieldrin, are closely related, dieldrin being the epoxide of aldrin.

Aldrin is 3,4,5,6,12,12-hexachlorotetracyclo $[6,2,1,1^{3,6},0^{2,7}]$ dodeca-4, 9-diene.

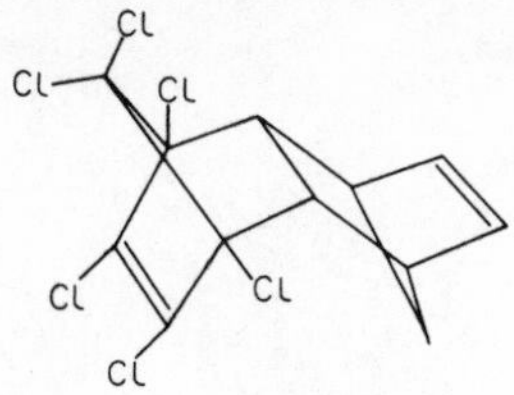

Spatial structure of Aldrin

XIV

This is made by the Diels-Alder reaction of the dienophile bicyclo(2,2,1)-hepta-2,5-diene with the diene hexachlorocyclopentadiene (XV).

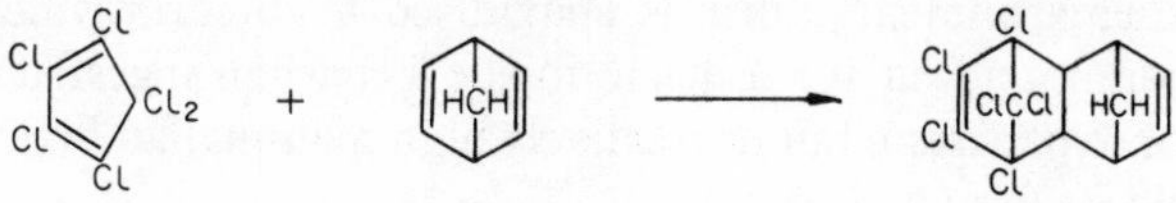

XV

As with most Diels-Alder reactions involving moderately reactive components the conditions are fairly mild. In this case it is only necessary to heat the diene with a molar excess of the dienophile at about 90 °C for about 18 hours and then to remove the excess of reactant by distillation under reduced pressure to obtain crude product which is easily purified by recrystallisation from methanol.

There are four stereoisomers of which aldrin is the only one to have been marketed successfully as an insecticide. However the epoxide of another stereoisomer has been used as an insecticide. This is endrin, which is isomeric with the more important dieldrin, the epoxide of aldrin. In both cases the epoxidation may be represented as shown (XVI).

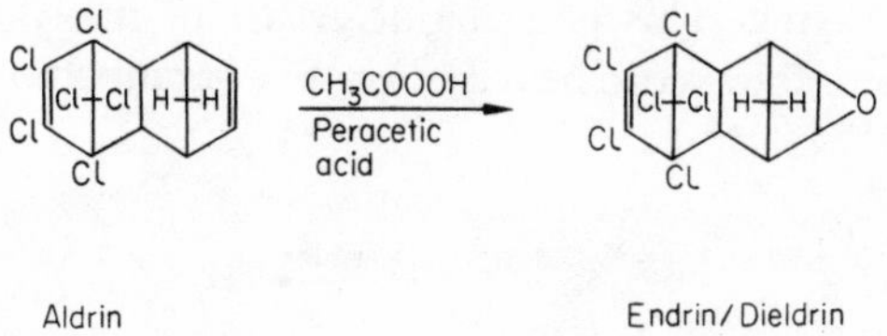

Aldrin   Endrin/Dieldrin

XVI

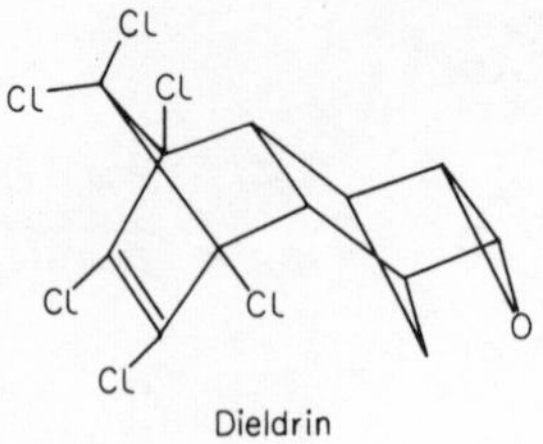

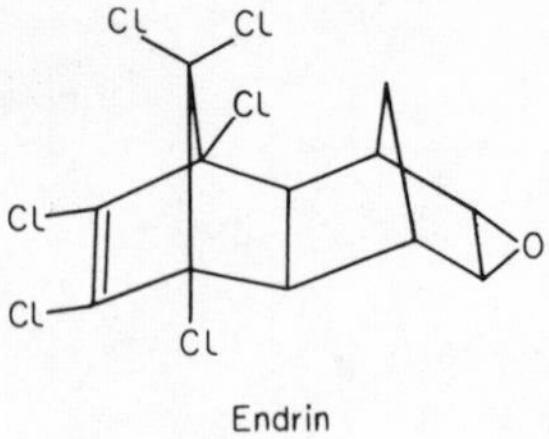

Dieldrin

Endrin

XVII   XVIII

These compounds have been used as seed dressings, soil insecticides, and in the case of endrin for controlling blackberry mite and blackcurrant gall mite. In the environment, aldrin is epoxidised to dieldrin which is itself a persistent compound having a deleterious effect when ingested by wildlife. Endrin is not so persistent but its relatively high mammalian toxicity requires that it should be used with care.

Although the organochlorine insecticides have been used very successfully and safely both in agriculture and in public health, the development of insect resistance to them which has necessitated the use of more active compounds to obtain the same effect, and the associated contamination of wildlife is resulting in their replacement by other agents, particularly the organophosphorus compounds.

Many of the organophosphorus compounds are not new. It is the discovery of their potent insecticidal activity which has made them so important. Unfortunately many are also very toxic to mammals so that special precautions have to be taken in their use.

One of the earliest useful organophosphorus insecticides is the ester tetraethylpyrophosphate (TEPP, XIX) which was synthesised by P. de Clermont in the nineteenth century but first marketed in Germany in 1943 as a substitute for nicotine in aphis control. It has a very high mammalian toxicity; in rats the acute oral LD_{50} is 1 mg/kg.

As with most organophosphorus esters, this compound is readily hydrolysed by water giving soluble products, so that it does not leave persistent residues after spraying. This is a characteristic of many members of this group, and is where they score heavily over the organochlorine compounds.

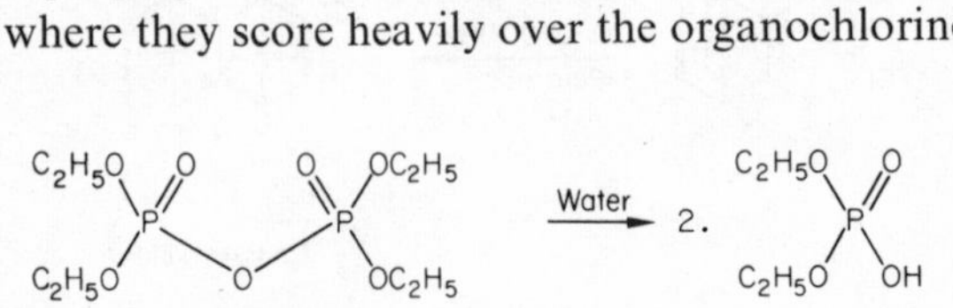

TEPP

XIX

Whereas TEPP is a contact insecticide, being similar in this respect to the organochlorine compounds, the corresponding dimethylamide, shradan (XX), has the interesting property of systemic activity; that is, it can be absorbed into the tissue of a plant and so make the plant itself lethal to insects which feed on it.

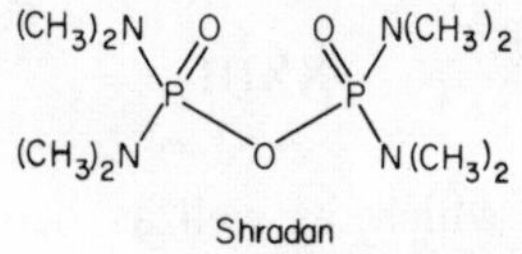

Shradan

XX

Although shradan is toxic to mammals systemic activity has been developed to a considerable extent in other insecticides which are less dangerous to use, and it points to the selective approach which is necessary if pests are to be destroyed without killing useful insects.

The mechanism of insecticidal activity is complicated but in the organophosphorus series there are clear indications of a common mode of action. Parathion (XXI) and related compounds even show a limited parallelism between changes in chemical structure and insecticidal activity.

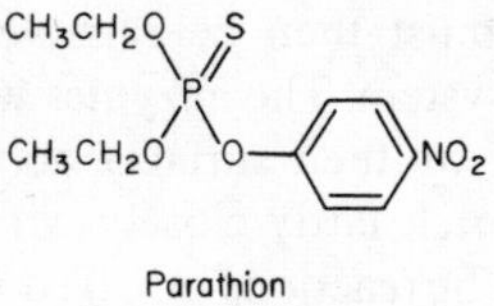

Parathion

XXI

A large number of compounds closely related to parathion have found use as insecticides and they can be synthesised by routes similar to that used for its preparation. The method used is simply an extension of one of the basic methods for the preparation of esters (XXII).

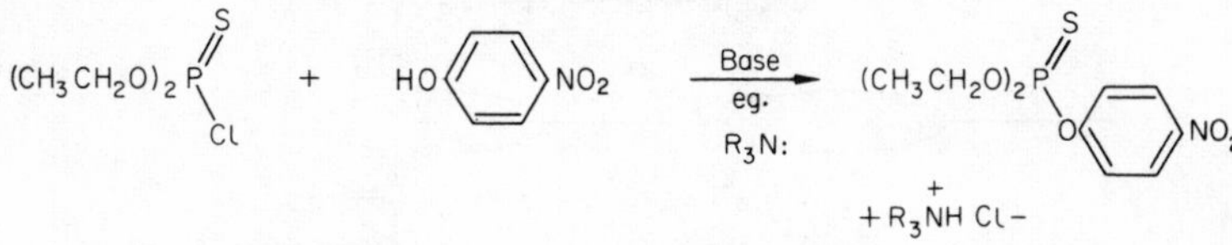

XXII

The acid chloride required for this reaction may be obtained from the reaction of alcohol and thiophosphoryl chloride ($PSCl_3$) or from phosphorus pentasulphide (XXIII).

$$P_2S_5 + 4\ CH_3CH_2OH \longrightarrow 2\ (CH_3CH_2O)_2P\overset{\displaystyle S}{\underset{\displaystyle SH}{\diagup\!\!\!\!\diagdown}} + H_2S$$

$$2\ (CH_3CH_2O)_2P\overset{\displaystyle S}{\underset{\displaystyle SH}{\diagup\!\!\!\!\diagdown}} + 3\ Cl_2 \longrightarrow 2\ (CH_3CH_2O)_2P\overset{\displaystyle S}{\underset{\displaystyle Cl}{\diagup\!\!\!\!\diagdown}} + 2HCl + S_2Cl_2$$

XXIII

Parathion is a liquid which, as well as being insecticidal, is toxic to mammals, especially by absorption through the skin. The corresponding methyl ester—methyl parathion—does not penetrate the skin so readily, presumably because the shorter alkyl chain length imparts lower fat solubility; also it is hydrolysed more easily. The toxicity to mammals as shown by the LD_{50} is only about a quarter of that of parathion so that, as one might expect, it is gradually replacing the latter as a commercial insecticide.

The insecticidal activity and mammalian toxicity of this group is closely connected with their cholinesterase inhibiting ability, although an examination of this alone gives only a simplified picture of their action. A more detailed treatment of this is given elsewhere;[8] but, very simply, in the nervous system impulses are transmitted at neuromuscular junctions by the liberation of acetylcholine which must then be removed by enzymic hydrolysis to prevent paralysis of the system. The enzymes which perform this hydrolysis, the cholinesterases, have on their surfaces active sites which are spaced in such a way that they match fairly closely with appropriate centres on the acetylcholine molecule. The reactions involved may be represented as follows (XXIV):

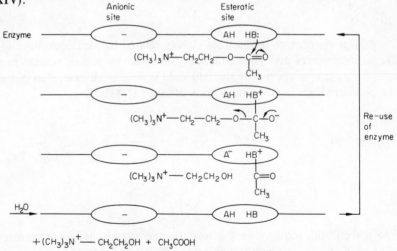

XXIV

Certain compounds in the organophosphorus and the carbamate groups have similar molecular dimensions so that they can compete with acetylcholine in this process.

The organophosphorus compounds phosphorylate the enzyme producing a stable entity which is no longer capable of hydrolysing the acetylcholine. As a result the nerve is paralysed and if sufficient centres are affected then death ensues. The reaction may be represented as follows (XXV):

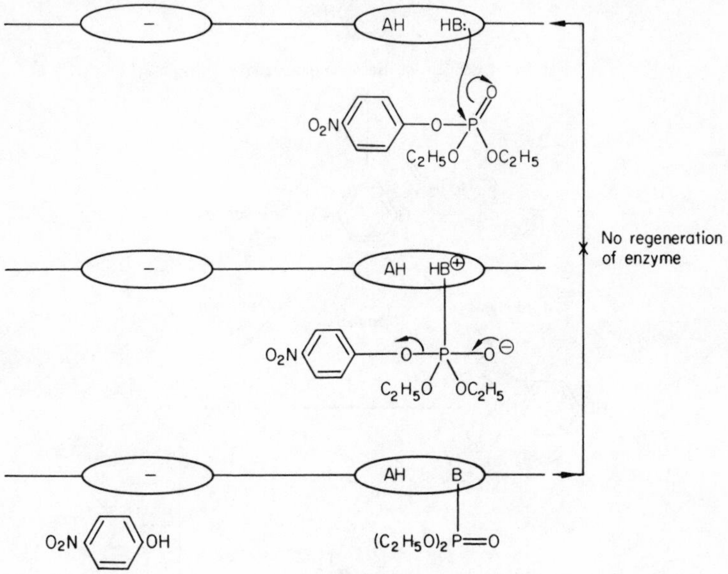

XXV

The effectiveness of this series of compounds in inhibiting cholinesterase in vitro, that is outside the living organism, is related directly to the weakness of organophosphorus bond which is broken in the process, and as might be expected this is enhanced by electron-withdrawing substituents in the benzene ring. In fact, so close is this connection that direct relationship can be shown between this activity and the acidity of the corresponding phenols.[9] However the situation inside the organism is somewhat more complicated so that the 2,4-dinitrophenyl compound which is highly active in vitro, is non-insecticidal, probably because the O-P bond is so weak that hydrolysis occurs in the insect before it can get to the normal site of action.

The process for transmission of nervous impulses is similar in mammals and insects so that one would expect to meet considerable danger in using these materials, but fortunately there are differences in metabolism which can be used to produce selectivity. In the previous example paraoxon was used to illustrate cholinesterase inhibition whereas the material used as an insecticide is its sulphur analogue, parathion. The insect is able to replace the sulphur by oxygen using an enzyme process and so convert the inactive parathion to the highly active paraoxon. In mammals various alternative degradative reactions are more important as shown in the following scheme (XXVI).

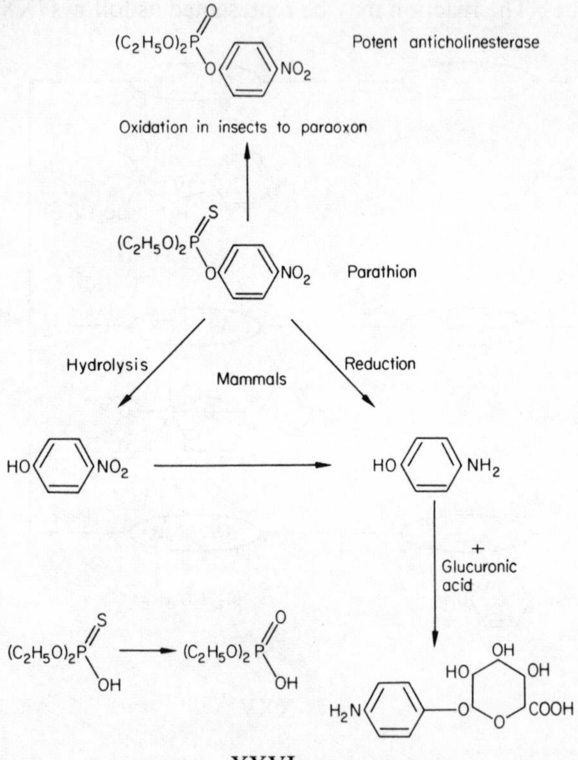

$(C_2H_5O)_2P$ ... NO_2 Potent anticholinesterase

Oxidation in insects to paraoxon

$(C_2H_5O)_2P$ ... NO_2 Parathion

Hydrolysis   Reduction

Mammals

HO ... NO_2 ——→ HO ... NH_2

$+$
Glucuronic acid

$(C_2H_5O)_2P$ ... OH —→ $(C_2H_5O)_2P$ ... OH

H_2N ... O ... OH HO OH $COOH$

XXVI

Degradation of many organophosphorus compounds in the environment occurs in a similar manner and in a short time giving ultimately phosphoric acids so that they do not pose long term problems of persistence.

The mammalian toxicity varies very considerably in the organophosphorus group from the highly poisonous TEPP to the virtually non-toxic fenchlorphos (XXVII) which is so safe with mammals that it may be used

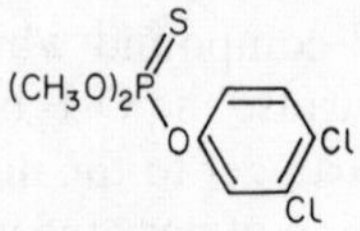

$(CH_3O)_2P$ ... O ... Cl Cl

Fenchlorphos

XXVII

internally in cattle for the systemic control of the warble fly which attacks their hides.

Whereas insecticides have a very general application to agriculture and public health, herbicides are used almost exclusively for the former, although considerable quantities have been used in defoliation of trees for military purposes; and in fact, the more notable toxicity problems in recent years stem from this gross use of herbicide. As can be seen from figure 3.1, of all the pesticides, the herbicide group is growing most rapidly, and already is at least as important as the insecticides. This reflects the importance of crop losses which are due to weeds, and possibly also the lesser importance of environmental problems. When compared with insecticides weeds are sufficiently different from mammals in their biological processes that the functions which are disrupted by herbicides do not have their exact counterparts in man. In consequence most modern herbicides are relatively non-toxic to mammals in general, and man in particular.

As with insecticides, some weed killers are very persistent. This ranges from a few years in the case of the triazine simazine when used as a total herbicide to the almost immediate adsorption of paraquat on normal soil which effectively removes it from further interaction with the environment.

The group of herbicides which paved the way for modern weed killing practice and the group which is still of greatest importance in volume of manufacture consists of the chlorophenoxy acids which have plant hormone-like properties.

Typical of this group and perhaps the best known is 2,4-Dichloro-phenoxyacetic acid (2,4-D; XXVIII).

$$Cl \overbrace{\bigcirc}^{OCH_2COOH} Cl$$

2,4–dichlorophenoxyacetic acid

XXVIII

This is a highly active and selective weedkiller which is used for the removal of broad leaf weeds in cereal crops, and more popularly in producing beautiful lawns.

The variety of related compounds which are active in this way is very large, but before considering them any further, it is interesting to show how they can be traced back to the early work on plant growth regulation. The discovery as early as 1928 of a compound isolated from urine which caused the stimulation of plant growth led to its identification as indole-3-acetic acid (XXIX) and subsequently to the preparation of many acetic acid derivatives which have this property.

A number of these are used for treating cuttings prior to planting to stimulate root growth, two examples being NAA (**XXX**) and NOXA (**XXXI**).

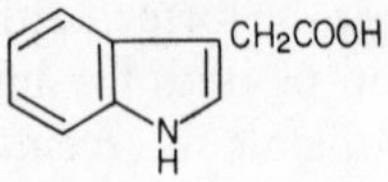

Indole – 3 – acetic acid	α – Naphthylacetic acid	β – Naphthoxyacetic acid
	NAA	NOXA
XXIX	**XXX**	**XXXI**

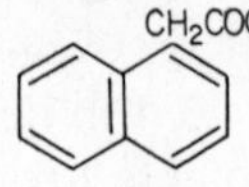

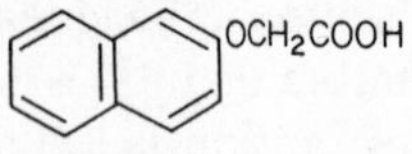

The discovery that cereals were unharmed by NAA but that it killed sugar beet was followed by tests on simpler molecules having similar features, with ultimately a herbicide bonanza in the substituted phenoxyacetic acids.

A common characteristic of these herbicides is that they stimulate growth, but this growth is excessive and disorganised so that tumours appear in the stems, leaves curl, the plant becomes grossly distorted and ultimately it dies.

These compounds have a significant effect only on broad leaf plants so that their major use is in the protection of cereal crops, the most important area of food production in agriculture.

Fortunately the raw materials for the manufacture of these compounds are very cheap and the reactions involved are simple and easy to control so that the cost of the product is generally very small compared with most other pesticides.

The preparation of 2,4-D is fairly typical of the whole series, the main reaction being the attack of a substituted phenoxide ion on the chloroacid salt (**XXXII**):

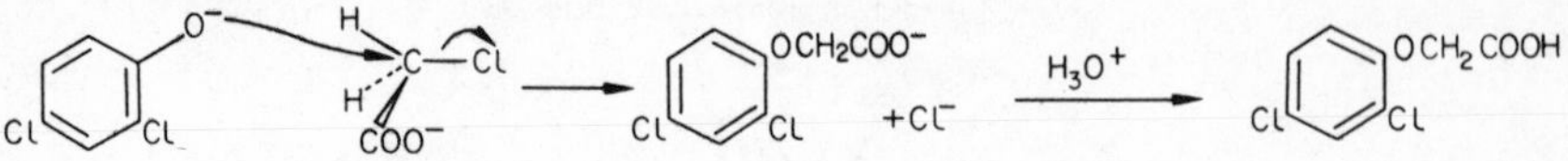

XXXII

The reaction is carried out with addition of sodium hydroxide solution to maintain a strongly alkaline medium so that as one might expect there is considerable hydrolysis of the chloroacetate to glycolate (**XXXIII**).

$$HO^- + ClCH_2COO^- \longrightarrow HOCH_2COO^- + Cl^-$$

XXXIII

Glycolic acid and its sodium salt are both very soluble in water so that they are easily removed from the 2,4-D acid which is only very sparingly soluble in water.

From an examination of spatial arrangement in 2,4-D one can expect chirality or optical isomerism to be introduced if one of the alpha hydrogens is replaced by an alkyl group (XXXIV):

2, 4-DP
2-(2, 4-dichlorophenoxy)
propionic acid

XXXIV

Biological systems are usually highly specific in their utilisation of chiral materials, one isomer only being acceptable. The herbicides are no exception and one finds that one enantiomorph is much more active than its fellow; however the products are made by simple chemical processes which will always give equal quantities of the two enantiomorphs and the expense of resolution is far too great, so that the commercial product is the mixture of isomers.

The free phenoxy acids have relatively low solubilities in water and are normally applied in agriculture as aqueous solutions of their salts. Sodium and potassium salts are obvious choices but some have rather low solubilities in water, and in general they do not form clear solutions quickly so that there is a risk of undissolved solids blocking the jets of agricultural spray equipment. Some organic salts on the other hand, such as those of ethanol-amine are very soluble in water, and these products are marketed as high concentration liquids which require only dilution with clear water prior to use.

An alternative form of the active compound which has been used successfully in agriculture is the ester, usually although not always of a long chain alcohol. This form which is more popular for aerial spraying has the advantage that it readily penetrates the fatty tissue of the plant where it is then hydrolysed to the anion, but at the same time it is only sparingly soluble and will not therefore be washed from the leaves so readily by rain.

A disadvantage, particularly of the esters of short chain alcohols is that they have sufficiently high vapour pressures that the vapour may contaminate crops in adjacent areas and add to the other disadvantage of aerial spraying which is the possibility of spray drift which can result in contamination of crops at a great distance from the point of application.

One compound which has been used extensively as its esters is 2-(2,4,5-trichlorophenoxy) acetic acid (2,4,5-T; XXXV).

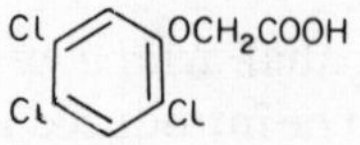

2, 4, 5 - T

2 - (2, 4, 5 - trichlorophenoxy)
acetic acid

XXXV

The ester from nonanol or a similar long chain alcohol, together with wetting and emulsifying agents, is dissolved in a hydrocarbon solvent such as kerosene. This product is effective in penetrating and killing woody plants such as bushes and small trees.

The impure technical compound has been used extensively as a defoliant for military purposes and as a result there has been the inevitable contamination of the local population which seems to have had a toxic effect, resulting in the production of abnormalities in unborn infants.[10] This is probably due not to 2,4,5-T itself but to an impurity Dioxin which is a by-product of its manufacture and is present in technical 2,4,5-T (XXXVI).

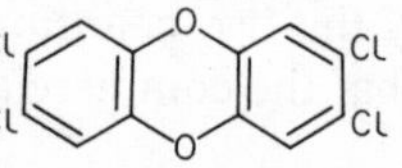

Dioxin

2, 3, 6, 7 - tetrachloro-p-dioxin

XXXVI

Toxicological tests have indicated that this compound is teratogenic (that is, causes abnormalities in the foetus). From an examination of the starting material for 2,4,5-T manufacture, that is 2,4,5-trichlorophenol, one can see that dioxin could be formed by nucleophilic aromatic substitution reactions, although the limited electron withdrawal by the para chlorine would indicate a fairly low reactivity (XXXVII).

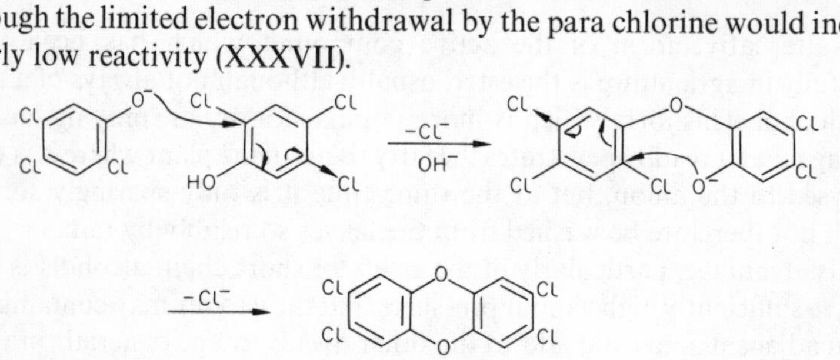

XXXVII

In spite of this unexpected toxicity, the group as a whole has proved useful. Modification gives rise to a large variety of active compounds, although they usually contain the basic structure of phenoxyacetic acid.

Selectivity may be introduced by utilising the β-oxidase system which is contained in some plants. For example, the compound XXXVIII will promote

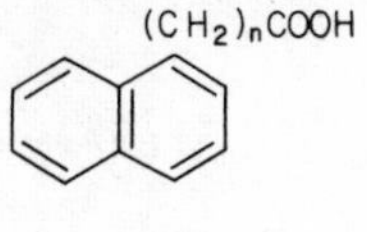

$$(CH_2)_n COOH$$

XXXVIII

the rooting of cuttings if n is odd, but not if it is even. This is because β-oxidase removes two carbons at a time (XXXIX).

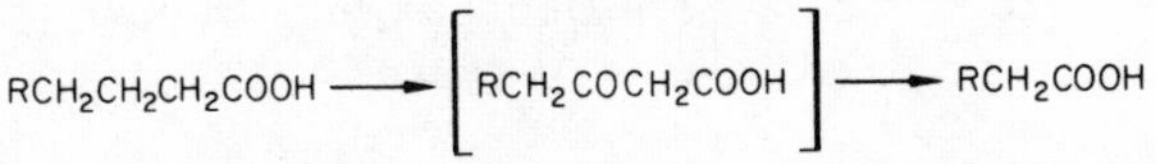

$$RCH_2CH_2CH_2COOH \longrightarrow \left[RCH_2COCH_2COOH \right] \longrightarrow RCH_2COOH$$

XXXIX

Only the system containing an odd value for n will be degraded to α-naphthylacetic acid which is the true growth stimulator. When n is even, then degradation will proceed as far as α-naphthoic acid which does not have this property.

This selectivity has been extended to herbicides by inserting two CH_2 groups into the side chain of the conventional compounds. One of these, MCPB (XL), has been marketed for controlling weeds in legume crops, relying on the fact that the latter cannot carry out β-oxidation efficiently, whilst the weeds have and are able to degrade it to the active herbicide MCPA (XLI).

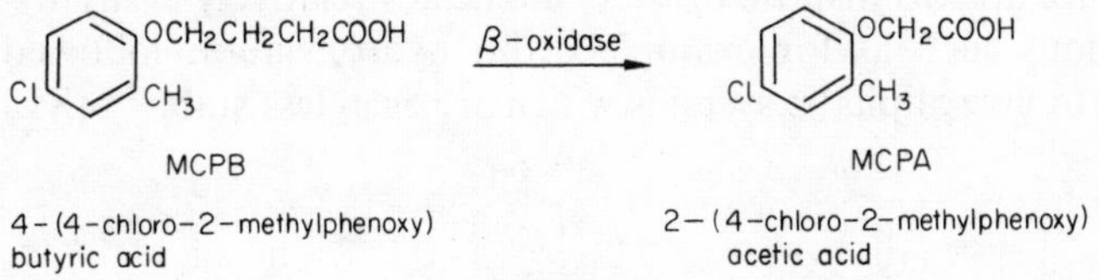

MCPB

4-(4-chloro-2-methylphenoxy)
butyric acid

MCPA

2-(4-chloro-2-methylphenoxy)
acetic acid

XL   **XLI**

Although these compounds are more expensive to manufacture than the corresponding aryloxyacetic acids, the methods are very similar (XLII).

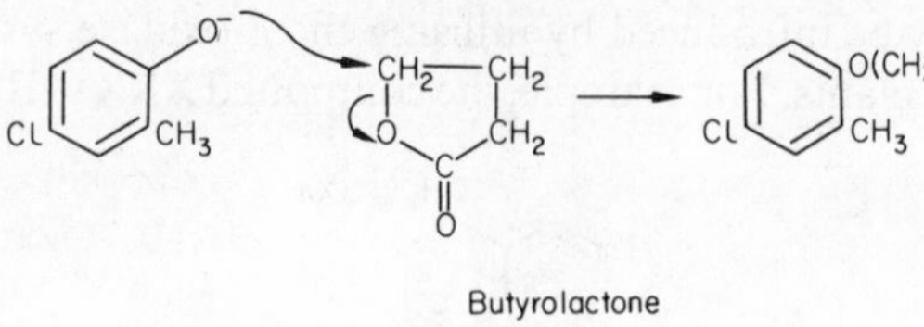

Butyrolactone

XLII

These examples by no means exhaust the tricks which can be played in this field. There have even been combinations of two herbicides suggested such as erbon which hydrolyses to the 2,2-dichloropropionate ion (XLIII),

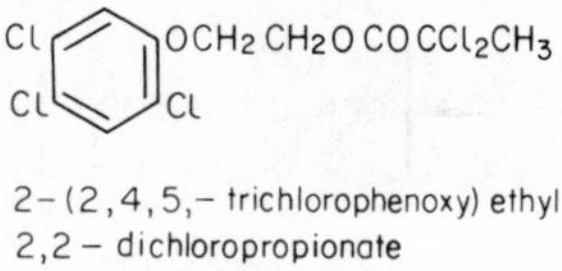

2-(2,4,5,- trichlorophenoxy) ethyl
2,2 - dichloropropionate

XLIII

one of the few herbicides active against couch grass; whilst the associated 2-(2,4,5-trichlorophenoxy) ethanol is oxidised by soil bacteria to the herbicide 2,4,5-T.

The substituted phenoxyacetic acids and related compounds form an extremely versatile and important group of herbicides which are used in great quantities, so that a consideration of their persistence and break-down in the environment is important. Unlike the organochlorine insecticides, these molecules are chemically highly reactive due to the carboxyl function which endows them with reactivity towards bases and hence water solubility. Also the alkoxyl group renders the aromatic nucleus reactive towards oxidation especially. It is not surprising therefore to find that biochemical systems in general are able to dispose of these molecules relatively easily by a variety of reactions such as for example ortho hydroxylation followed by ring opening to give aliphatic systems which are even less stable[11] (XLIV).

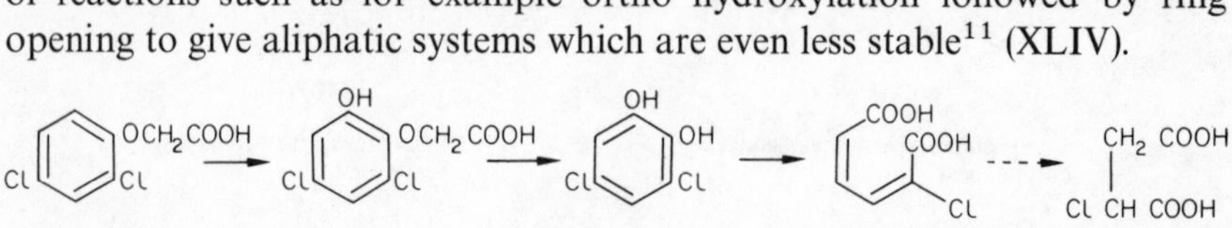

XLIV

On the whole, one can expect that these compounds will not persist for years in the environment. In mammals the breakdown may not be so complete as in the soil, but the herbicides themselves and their metabolites are fairly soluble compounds which will be excreted readily.

Most of the other herbicides in use today contain nitrogen, a particularly important group being the triazines, one of which—atrazine—has the highest sales amongst all pesticides in the USA.[12] These are derivatives of cyanuric chloride, a product of the polymerisation of cyanogen chloride in the presence of a catalyst such as anhydrous aluminium chloride (XLV).

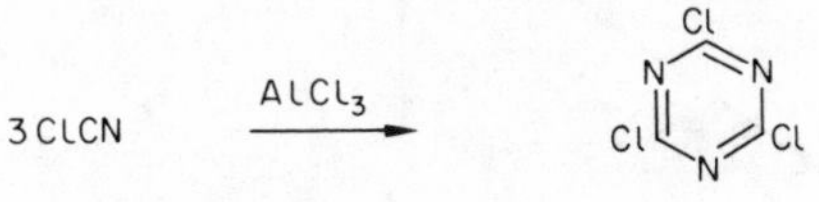

$$3\,CLCN \xrightarrow{ALCL_3}$$

Cyanuric chloride 2, 4, 6–
trichloro–s–triazine

XLV

The three chlorine atoms are reactive towards nucleophiles and they can be replaced stepwise by, for example, amines, giving the possibility of a variety of simple derivatives, many of which have herbicidal activity (XLVI).

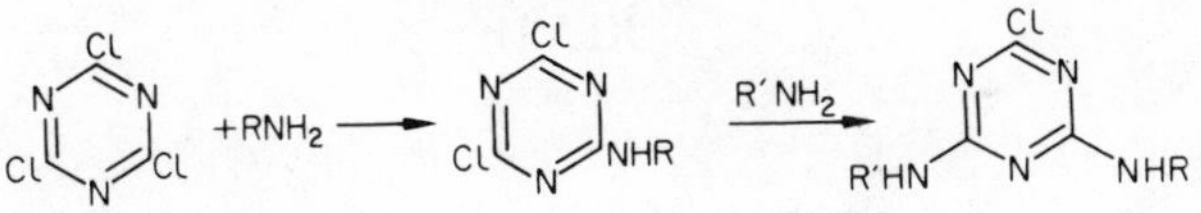

e.g.	R	R'	Herbicide
	$-C_2H_5$	$-C_2H_5$	Simazine
	$-C_2H_5$	$-CH(CH_3)_2$	Atrazine
	$-CH(CH_3)_2$	$-CH(CH_3)_2$	Propazine

XLVI

The methyl ethers and the thio ethers which are obtained by replacement of the remaining chlorine atom are also herbicidal.

Both simazine and atrazine are used as herbicides in maize crops, but the former has an important application at high dosage as a persistent non-selective weed killer on the roadside and on railways, etc.

The mode of action of these compounds appears to be quite different from the hormone herbicides. In place of the growth distortion which is obtained with the latter, the triazines cause the leaves to yellow and dry, followed by death of the plant. Simazine inhibits the accumulation of starch, indicating that it acts by interfering with the plant's photosynthetic process. In those plants which are unaffected, it appears that the triazines readily undergo hydrolysis and dealkylation. Degradation in the soil and in mammals gives rise to similar intermediates (XLVII).

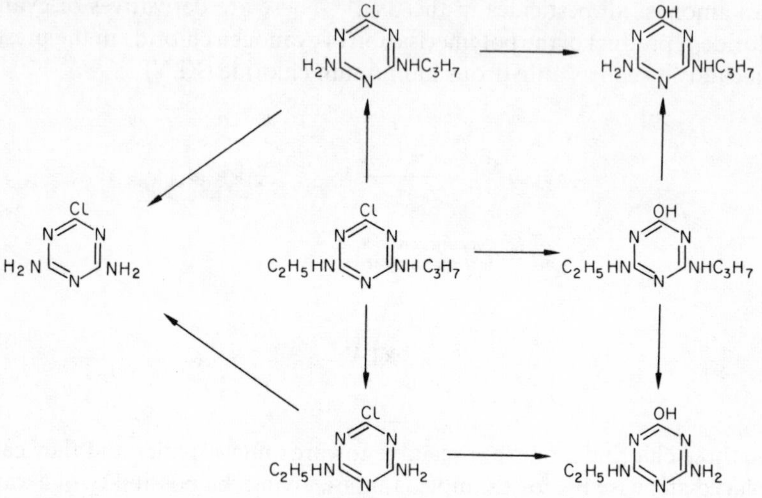

XLVII

The more notable members of this series of herbicides have a very low mammalian toxicity. In rats the LD_{50} is in the region of 5,000 mg/kg. The almost complete lack of environmental problems with the members of this group augurs well for their future.

A highly successful, although rather expensive, nitrogen-containing herbicide is paraquat, the most important member of the bipyridyl series of compounds[13] (XLVIII).

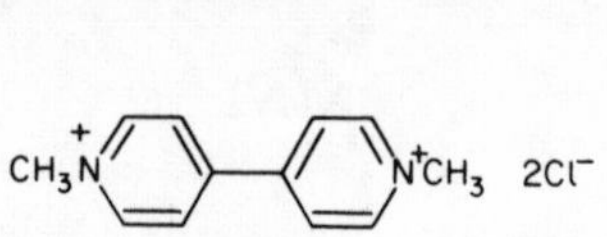

Paraquat

XLVIII

This compound has the advantage of rapid absorption through plant leaf tissue, but also rapid deactivation on soil by more or less permanent adsorption. In grasses on which it is most active the top growth is killed in a very short time, especially in sunlight, suggesting as in the case of the triazines that interruption of photosynthesis is occurring.

The mechanism of its action is related to the ease of production of a radical for which several canonical forms can be drawn and which is therefore relatively stable (XLIX).

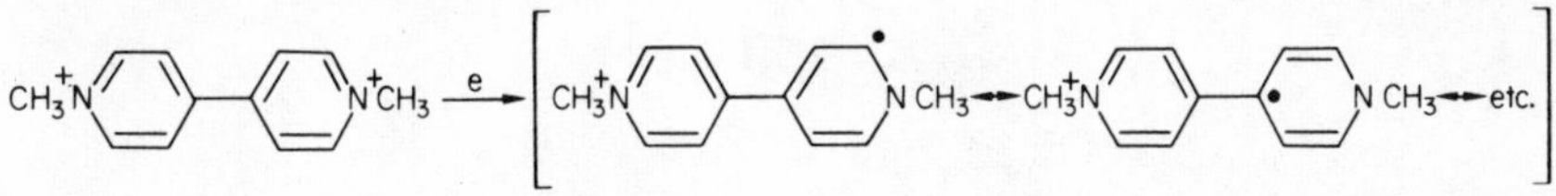

$$CH_3\overset{+}{N}\langle\rangle\langle\rangle\overset{+}{N}CH_3 \xrightarrow{e} \left[CH_3\overset{+}{N}\langle\rangle\langle\rangle\overset{\bullet}{N}CH_3 \longrightarrow CH_3\overset{+}{N}\langle\rangle\langle\rangle\overset{\bullet}{N}CH_3 \longrightarrow etc. \right]$$

XLIX

The formation of this radical can easily be demonstrated by shaking a solution of paraquat with a reducing agent such as zinc dust when an intense blue develops which fades in contact with air.

In the group of related compounds, it has been shown that their herbicidal activity parallels to a certain extent the redox potential of the compounds, which would be the case if the radical were involved in the physiological mechanism.

It has been suggested that in the plant the radical is involved in the generation of hydrogen peroxide from water and oxygen (L) and that this is the true toxicant where $R°$ is the radical from paraquat. This however is an oversimplification and the mechanism is more complicated, but it is fairly certain that the radicals play an important part in the overall process.

$$R^{\bullet} + O_2 \longrightarrow R-OO^{\bullet} \xrightarrow{H_3O^+} R^+ + HOOH$$

L

This type of weedkiller with its unique character offers the possibility of novel types of cultivation so that for example land has been seeded straight after paraquat treatment without ploughing. The results have been very encouraging and will probably pay dividends, particularly in areas where soil erosion is a serious problem since the roots remaining after treatment with weedkiller will tend to hold the soil together.

The environmental problems, so far as soil and plants are concerned, are minimal. In the former case, adsorption effectively removes the material permanently, whereas in non-susceptible plants degradation occurs via ring opening to the relatively non-toxic quaternary salt of isonicotinic acid (LI).

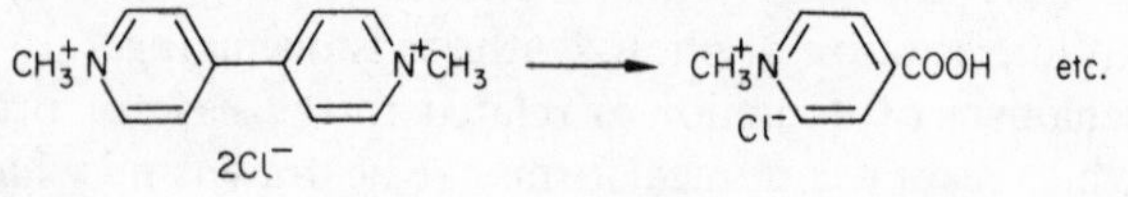

$$CH_3^+N \langle\rangle - \langle\rangle N^+CH_3 \quad 2Cl^- \longrightarrow CH_3^+N \langle\rangle COOH \quad Cl^- \quad \text{etc.}$$

LI

In man there is considerable danger if the substance is swallowed—the result is often fatal—so that the storage of the more concentrated solutions used in agriculture has to be carefully supervised. The compound is not readily absorbed through the skin, so that apart from direct ingestion into the stomach it is a safe herbicide to use.

So far only a selection of the pesticides which have been used on a large scale have been discussed since they are the ones most likely to lead to environmental problems. However, other substances are being developed which in the long term might offer more specific ways of dealing with pests. Mention should be made of the sex attractants, food lures and other agents which are capable in very small concentrations of affecting the behaviour of insects. They are generally fairly complex molecules and in the main both difficult and expensive to synthesise. For example, the female gypsy moth produces a male attractant which has been isolated and characterised. The natural compound has the cis configuration and is so active that the presence of only a few molecules is sufficient to attract the male (LII).

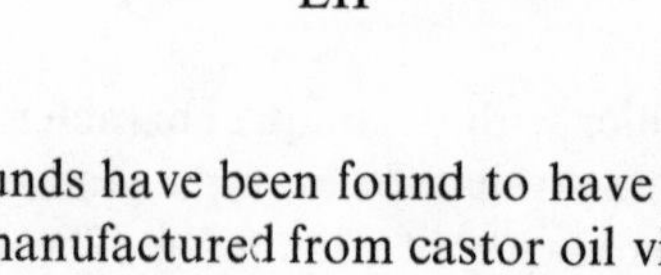

$$\begin{array}{c} \quad H \qquad\quad H \\ \quad \backslash \qquad\quad / \\ \quad C === C \\ CH_3(CH_2)_5CHCH_2 \qquad (CH_2)_5CH_2OH \\ \qquad | \\ \qquad OCOCH_3 \end{array}$$

Natural gypsy moth attractant

LII

Analogous compounds have been found to have similar activity, one of which gyplure can be manufactured from castor oil via ricinoleic acid (LIII).

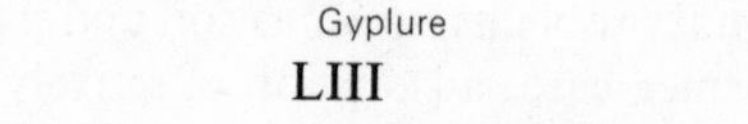

$$CH_3(CH_2)_5CHCH_2CH == CH(CH_2)_7CH_2OH$$
$$\qquad\qquad | $$
$$\qquad\qquad OCOCH_3$$

Gyplure

LIII

The use of this type of compound in conjunction with a poison or sterilant could give a more selective insect control; on the other hand there has been some success in using this type of compound in traps to monitor the infestation which is occurring prior to the application of pesticides.

In general, these substances—since they are more specific—will have only limited markets. When one considers that launching a new product now costs about five million dollars and that future government requirements for toxicity testing will be even more stringent than at present, then one can well imagine that products having a narrow spectrum of activity and therefore a limited market are unlikely to come from industry unless there is government support for their development and manufacture.

Looking at the situation from a global point of view, it is probable that the population at large is well served in the short term by pesticides, in spite of their environmental hazards. Let us hope that the greater participation by government agencies, which seems to be inevitable if we are to move away from our present rather crude forms of pest control, will not discourage the enthusiastic contribution from industrial research which has brought us so far so quickly.

References

1. H. H. Cramer (1967). *Plant Protection and World Crop Production*, Bayer
2. *Chem. and Eng. News* (1970). Sept. 7, 76A.
3. Business Monitor Production Series, Pesticides and Allied Products (1970), HMSO, London
4. E. R. Laws and F. J. Biros (1967). *Arch. Environ. Health*, **15**, 766
5. *Nature* (1972), **237**, 420, 422–424, Editorial comment
6. F. W. Whittemore (1970). *Technological Economies of Crop Protection and Pest Control*, S.C.I. Monograph No. 36, 230–234.
7. R. G. Nash and E. A. Woolson (1967). *Science*, **157**, 924
8. R. W. Riseborough *et al.* (1968). *Science*, **159**, 1233
9. T. R. Fukuto, 'The Chemistry and Action of Organic Phosphorus Insecticides'. *Advances in Pest Control Research* (Vol. 1), 147–192
10. K. D. Courtney *et al.* (1970). *Science*, **168**, 864
11. M. A. Loos (1971). Pesticide Terminal Residues, *I.U.P.A.C.*, 291–303
12. J. Neumeyer, D. Gibbons and H. Trask (1969). *Chem. Week.*, Pt. 1, April 12
13. A. Calderbank. *Advances in Pest Control Research* (Vol. VIII) (1968)

CHAPTER 4

AIR POLLUTION

R. PERRY (Public Health Engineering)

and

D. H. SLATER (Chemical Engineering)
Imperial College

The atmosphere can be considered as a source of continuous chemical reaction. It readily absorbs a range of solids, liquids and gases from both natural and industrial sources. These can then disperse, react among themselves, or react with other substances already present in the atmosphere. These substances or the products from their reactions finally find their way into a sink, such as the ocean, or reach a receptor such as man. The solids and liquids are usually dispersed in the atmosphere in the form of aerosols. It is difficult to define the term pollutant, but generally speaking, any substance that ultimately causes annoyance or discomfort or danger to man can be included in this category. This means, of course, that the concentrations at which these materials become pollutants must be defined and also critical concentrations of possible precursors evaluated. Obviously then the total weight of a particular pollutant produced is only of secondary importance to its effect upon the environment.

The actual concentration of pollutants in the air owes as much to dispersion mechanisms as to their production and removal. Some of the lighter gases released into the air (e.g. Helium) can escape from the earth's atmosphere, while others such as carbon dioxide enter the atmosphere faster than the sink can remove them, thus actually accumulating as an increasing background.

Normally the atmosphere itself disperses the pollutant by mixing it thoroughly with a very large volume of air, diluting the pollution to acceptable levels. The rates of dispersion vary widely with the topography and meteorological conditions. An example of an inversion process is shown in figure 4.1 in which dispersion by convection and transport by winds becomes impossible. As the earth's surface becomes warmed by sunlight, the layer of air in contact with ground is also heated by conduction. This warmer air is less dense then the cold air above it and rises to produce convection

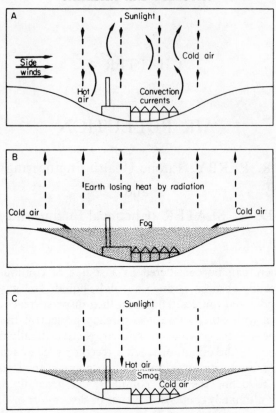

Figure 4.1:   Inversion (effect of topography and weather)

currents. Thus pollutants produced in the surface layer are effectively dispersed. On a still evening, the process is reversed, as the earth cools, producing a static layer of cold air in the absence of mixing due to winds. If this in turn induces condensation or fog, the morning sun cannot initially penetrate the fog layer now liberally laced with urban emissions so that the solar heating cycle cannot establish itself. The well of cold air trapped by the layer of warmer air now becomes a closed chemical reactor in which all the products of an urban society, traffic, industrial and domestic emissions, are trapped and build up to abnormally high concentrations.

In order to assess the levels at which different pollutants exist it is necessary to use specific and sensitive analytical techniques to measure their concentrations. Substances being measured are often only present at concentrations in the order of micrograms per cubic metre (μg/m^3), but it is often at this level that many primary pollutants can initiate a photochemical reaction chain that can lead to large concentrations of more undesirable materials.

Frequently an extended monitoring period is required before adverse time/concentration effects of a particular pollutant can be demonstrated. The source or sources have to be clearly determined and the dispersion under differing meteorological conditions has to be established. Even at this stage it is difficult to realistically assess the hazards of many substances. Although the toxicity of individual chemicals may well be understood at higher concentrations, the effect of long exposure to small concentrations of mixtures is more uncertain.

When the build up of a polluting material in the atmosphere has been demonstrated then the techniques available for control must be reviewed. These range from the use of alternative fuels to more rigid clean-up of effluent gases. The choice of method used is determined by several factors and can be largely influenced by economic considerations according to the standard that is required. Obviously there is no point in setting a standard based upon the use of, say, a low sulphur content fuel, where nationally there is insufficient of this fuel to meet the requirements under periods of heavy demand. Once it has been shown, however, that control is feasible and desirable the last stage in the process is the introduction of legislation to ensure that predetermined standards are met.

The chemistry of nitrogen oxides plays an important part in air pollution. The equilibrium established at high temperatures in internal combustion engines, $N_2 + O_2 \rightleftharpoons 2NO$, is frozen by the expansion cycle. The nitric oxide so produced is then emitted in the exhaust gases, at locally high concentrations, that are high enough to make the subsequent atmospheric oxidation

$$2NO + O_2 \longrightarrow 2NO_2$$

I

relatively efficient. This nitrogen dioxide is a very efficient ultra violet absorber and is visible in the brown haze characteristic of photochemical smog.

$$NO_2 + h\nu \longrightarrow NO + O$$

II

The photodissociation produces highly reactive oxygen atoms which initiate the complex sequence of reactions leading to the formation of eye irritating components such as ozone, aldehydes, peroxyacyl nitrates (for example $CH_3 . COO . ONO_2$) and alkyl nitrates.

$$O + O_2 + M \longrightarrow O_3 + M$$

III

In polluted atmospheres the oxygen atoms and ozone (the 'oxidant'), can react further with other constituents, that is hydrocarbons, especially alkenes and alkyl benzenes, sulphur dioxide, carbon monoxide and particulates.

The photolysis of nitrogen dioxide and the formation of ozone are examples of chemical reactions involving not just 'stable' molecules but highly reactive fragments of molecules, in this case oxygen atoms.

The nitrogen dioxide molecule will absorb a photon of visible light (of wavelength less than 430 nm) and the energy of the absorbed photon ($= h\upsilon$) produces an 'excited' molecule of nitrogen dioxide which is much less stable than the normal 'ground' state and falls apart.

The occurrence of photochemical smog is due to the fact that these reactive fragments (known as 'free radicals') and 'excited' states of molecules are produced in an atmosphere containing hydrocarbons.

The hydrocarbons are then oxidised by these reactive species to form the peroxides and peroxy derivatives characteristic of smog atmospheres.

A simplified mechanism can be outlined, although the exact mechanism is the subject of much debate especially the efficient conversion of the nitric oxide to the absorbing nitrogen dioxide.

$$NO + O_2 \longrightarrow NO_2$$

IV

Once this is formed we can easily set up a series of reactions to keep the nitrogen dioxide in equilibrium.

$$NO_2 + h\upsilon \longrightarrow NO + O$$

V

$$O + O_2 \longrightarrow O_3 \qquad\qquad O_3 + NO \longrightarrow NO_2 + O_2$$

VI   VII

Thus the reaction is effectively photosensitised by the nitrogen dioxide.

Fragments of hydrocarbons can be produced by oxygen atom attack. If we split methane, CH_4, into a hydrogen atom H, and a neutral fragment CH_3, we have effectively shared the paired electrons in the C-H covalent bond between the two fragments. The CH_3 thus contains an unpaired, extra electron and is known as the methyl free radical and can be represented by CH_3 , where the dot indicates the odd electron.

In general the parent hydrocarbon RH splits into fragments or free radicals when heated or attacked by other atoms or free radicals. If the radical has less carbon atoms than the parent we can indicate this with a superscript prime.

These free radicals are so much more reactive than stable molecules that they give rise to sequence of reactions which have the characteristic of forming products but maintaining the free radical concentration. These are known as free radical 'chain' reactions of which combustion is the most common example. In fact photochemical smog can be thought of as the low temperature combustion of hydrocarbons in the atmosphere—a nitrogen dioxide photosensitised free radical chain oxidation.

These chain reactions have three distinct phases.

(1) Initiation, where the free radicals are produced.
(2) Propagation, where stable products are formed and more free radicals produced.
(3) Termination, where the free radicals are removed from the system by reactions.

A possible set of reactions could be:

$$\text{Chain initiation} \quad RH + O \longrightarrow R'O\cdot + R''CHO$$

VIII

$$\text{Chain propagation} \quad R\cdot + O_2 \longrightarrow R'O_2^-$$

IX

$$R'O_2^- + NO \longrightarrow R'O\cdot + NO_2$$

X

$$R'O\cdot \longrightarrow R''\cdot + \text{Product}$$

XI

$$\text{Chain termination} \quad R'O\cdot + NO_2 \longrightarrow R'ONO_2$$

XII

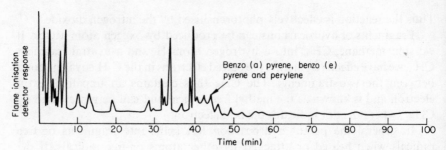

Figure 4.2:   GC soot extract

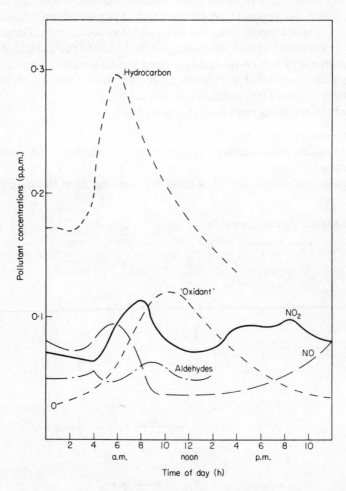

Figure 4.3:   Typical variation of pollutant concentrations during a 24-hour period

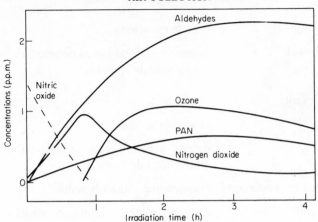

Figure 4.4: Products of the UV photolysis of vehicle emissions

The free radicals generated can then react further to produce oxidation products and compounds such as PAN, peroxyacyl nitrates, RCO_3NO_2, which cause intense eye irritation and bronchial problems.

This simplified treatment at least allows us to appreciate the qualitative behaviour of pollutant variation observed in smog (see figure 4.3). The extrapolation to a simulated smog chamber where the ultra violet photolysis of a synthetic mixture of exhaust emissions under controlled conditions shows even more clearly that we have got the essential features right (see figure 4.4).

Types of pollutant and their sources

The following is a brief list of general and specific air pollutants; and their most probable sources.

Common pollutants

Sulphur dioxide	domestic fires, electricity generation, oil refineries, brickworks, iron and steel works.
Particulates	vehicle emissions, domestic fires, gas works, electricity generation, incinerators, cement works, oil refineries, lime kilns, ceramic manufacturers, foundries, coke ovens.
Hydrocarbons	vehicle emissions, petroleum refineries.
Oxides of nitrogen	vehicle emissions, nitric acid plants, electricity generation, iron and steel works, fertiliser plants.

D

Carbon monoxide	vehicle emissions.
Carbon dioxide	domestic and industrial combustion of fuels, motor vehicle emissions.

Specific pollutants

Ammonia	ammonia works.
Bromides	motor vehicle emissions (from ethylene dibromide additives).
Chlorinated hydrocarbons	dry-cleaning establishments.
Chlorine and hydrogen chloride	chlorine works, aluminium works, metal recovery plant, refuse incinerators.
Fluorine and fluorides	brickworks, glassworks, aluminium smelting.
Mercaptans	oil refineries.
Metals	specific to appropriate works.
Sulphides	electricity generation, metal smelting, rubber vulcanising, coke ovens.

It is estimated that on a world wide basis, as much as 80 per cent of the sulphur dioxide in the air at any time is originally emitted as hydrogen sulphide which is then subsequently converted into sulphur dioxide. Of the remaining 20 per cent emitted as sulphur dioxide itself, 16 per cent comes from the combustion of sulphur-containing fuels, whereas processes like petroleum refining account for the remaining 4 per cent. In urban surroundings, however, the sulphur dioxide levels can be related to the sulphur content of fuels.

Particulates can be chemically very complex both in their reactions among themselves and in their interactions with other liquids and gases. Those up to 10 microns in diameter are the most numerous in the atmosphere and many of them contain liquids and gases adsorbed onto their surfaces, often arising from the same combustion process as the particulate itself. Figure 4.2 illustrates a gas chromatographic separation of several of these organic compounds separated from soot particles. Included are carcinogenic compounds like benzo(a) pyrene which are normally products of incomplete combustion and these, it has been claimed, could be a possible cause of lung cancer. The removal of particles from air largely depends upon size, so many of the smaller particles need to coagulate due to collision, before settlement can occur. Alternatively removal can take place following impacts with buildings or other objects.

It has been suggested that both carbon dioxide, also arising from combustion sources, and particulates could affect global temperatures. Carbon dioxide has the ability to absorb infra red radiation emitted by the earth and radiate it back to the earth while the shorter incident wavelengths from the sun readily pass through the atmosphere. This is known as the greenhouse effect. It is estimated that an increase in carbon dioxide content of 10 per cent would increase global temperatures by 0.5 °C. Opposing this effect is the increase in aerosols and particulates causing scattering of incident radiation from the sun and hence a lowering of temperature.

The sole significant man-made source of carbon monoxide is incomplete combustion and it is estimated that road traffic contributes more than 80 per cent of the total figure. Under normal conditions it is unreactive and escapes into the general atmosphere where it contributes about 10 per cent to the natural sources. In large cities over 95 per cent is man-made and levels can be 50–100 times higher than levels taken at remote sampling points which indicate global background levels of 0.1 ppm. The main sink, probably biological, has not been positively identified as yet, although the leaves of plants have been shown to be efficient in converting CO to CO_2.

Other pollutants from combustion sources are hydrocarbons and oxides of nitrogen. Of the total hydrocarbon emissions, man produces an estimated 15 per cent, mainly from incomplete combustion of hydrocarbon fuels and their refining. Natural sources include forests and vegetation (terpenes) and anaerobic bacterial processes (methane).

The oxides of nitrogen present in the atmosphere are mostly nitrous oxide, nitric oxide and nitrogen dioxide. Nitrous oxide occurs naturally and has a concentration of about 0.25 ppm. As the formation of nitric oxide from a mixture of nitrogen and oxygen is thermodynamically favoured at high temperatures, then processes which cause intense local heating, such as combustion, lightning, etc., can produce nitric oxide as an inadvertent by-product. Nitrogen dioxide can be produced by biological processes in the soil and by atmospheric oxidation of nitric oxide.

Sources of specific pollutants are readily identifiable and given in the table.

Urban chemistry

The oxidation of nitric oxide to nitrogen dioxide in the atmosphere is a relatively simple example of the subsequent reactions which occur when pollutants are released into the urban environment and interact.

Sulphur dioxide, implicated in the notorious 'killer' London smogs, probably causes bronchial damage through its oxidation product sulphur trioxide. This forms droplets of sulphuric acid in moist air. The exact mechanism of the oxidation is still uncertain but it is catalysed by metal ions present in particulates and the effect of sunlight on vehicle emissions. Thus small amounts of sulphur dioxide are probably harmless in the absence of

.... sunlight, but become lethal in polluted environments. .... drocarbons depends upon their reactivity or degree

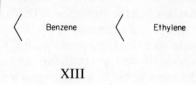

Benzene　　　　Ethylene

XIII

...ess than 1 ppm of reactive hydrocarbons and less then ....gen are required.

In the United Kingdom, high ozone levels which vary in step with sunlight intensity have been recently observed. It looks very much as if the favourable topography and much lower incidence of inversions in the major cities such as London inhibit the formation of this type of photochemical smog. The Clean Air Act has also done much to reduce the levels of particulates and sulphur dioxide responsible for London-type smogs. Nitrogen oxides emitted by supersonic air liners flying near the ozone layer may produce a local disturbance of this essential far ultra-violet filter by the reaction

$$NO + O_3 \longrightarrow NO_2 + O_2$$

XIV

The ozone layer should however quickly re-establish itself at a lower level, but as the exact mechanism of its formation is uncertain a categorical assurance that there will be no perturbation of the atmospheric structure at high altitudes cannot be assumed.

Vehicle emissions and their control

From a consideration of atmospheric chemistry, the four main undesirable materials emitted from vehicle exhausts, are carbon monoxide, hydrocarbons, oxides of nitrogen and particulates.

If hydrocarbons are to be efficiently combusted, an excess of air is required. The products would then contain predominantly carbon dioxide and water, together with a smaller quantity of nitrogen oxides formed by combination between oxygen and nitrogen within the engine. The effects of a variable air to fuel ratio can be seen from figure 4.5. Although a high air to fuel ratio gives the best economy and the most efficient combustion, this is not necessarily compatible with the best performance and some compromise is necessary. The requirements obviously vary and depend upon the mode of operation of the car (idling, cruising, accelerating or decelerating) but in general fuel/air ratios of about $\frac{1}{15}$ are required for complete combustion. In practice richer mixtures are often used leading to increased emission of products of incomplete combustion including carbon monoxide, unburnt hydrocarbons, aldehydes, ketones and other organic compounds. The emission of oxides of nitrogen tends to be highest at the stoichiometric air/fuel ratio corresponding to the highest temperature in the combustion zone, a factor that causes considerable difficulty when trying to institute control of vehicle emissions. Thus again some compromise is necessary, for if hydrocarbon emissions are minimised then high emissions of nitrogen oxides occur.

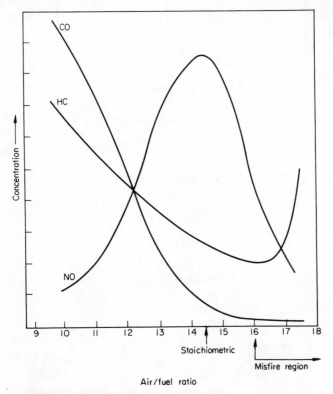

Figure 4.5:   HO, CO, NO emissions relative to mixture strength

A further factor, related to engine load, that influences the composition of exhaust gases is the manifold vacuum. This ranges from 0 inches of mercury during maximum acceleration to 25.2 inches during deceleration where, although the volume of exhaust gases decreases, there is a large increase in percentage of unburnt fuel. The concentration of the various exhaust pollutants in each of these modes is therefore illustrated in figure 4.6.

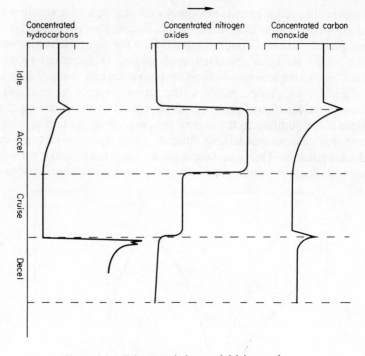

Figure 4.6:   Exhaust emissions and driving mode

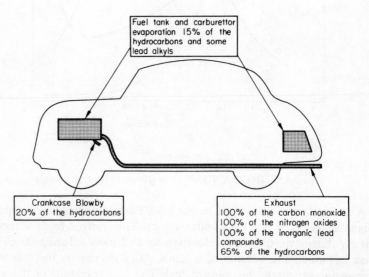

Figure 4.7:   Emissions from an uncontrolled vehicle

It can be seen here again that the formation of the nitrogen oxides occurs in a pattern that is the reverse to that for hydrocarbons. Concentrations of nitric oxides are highest during acceleration and low during deceleration and they increase as the air/fuel ratio increases. It must be remembered of course that the total volume of exhaust gases varies considerably during these modes of operation. This ranges from 5 cu. ft./min. during idling and deceleration up to 200 cu. ft./min. during acceleration.

Figure 4.7 illustrates the main sources of pollution from an uncontrolled car. Apart from the obvious exhaust emissions, these include evaporative losses from the carburettor and fuel tank and blow-by gases which are the gases that escape around the piston rings, particularly on the compression stroke. Of the hydrocarbons emitted by such a vehicle the exahust gases account for about 65 per cent, evaporative losses account for 15 per cent and the remaining 20 per cent are lost in the blow-by gases. Carbon monoxide, nitrogen oxides and particulates are emitted primarily from the exhaust.

If no controls were imposed, a continued increase in all pollutant levels could be expected directly related to the increase of vehicles on the roads (see figure 4.8). In these circumstances levels could be achieved at which

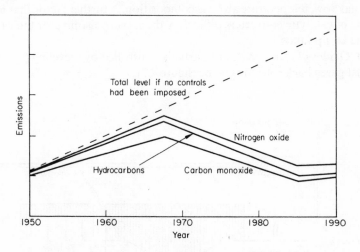

Figure 4.8: Predicted trends in vehicle emissions following introduction of controls

carbon monoxide would be definitely toxic and in countries enjoying long periods of sunshine the hydrocarbons and oxides of nitrogen would lead to considerable formation of photochemical smog. Again it would be undesirable to permit an unrestricted increase in particulate emissions as they contain a large proportion of the inorganic lead compounds and some carcinogenic materials. The following methods are available for controlling these emissions:

(1) Exhaust emissions can be controlled to some extent by optimising on air-fuel ratios, ignition timing and maintenance and cylinder design.

(2) Crank case blow-by gases can be recycled into the engine intake. This is now becoming standard practice to seal the crank case and recycle the gases for recombustion.

(3) Evaporative losses can be reduced by absorbing the hydrocarbons into a charcoal filter.

(4) Exhaust emissions of hydrocarbons and carbon monoxide can be reduced by air injection into the manifold by the exhaust valves. This ensures that further oxidation takes place prior to the gases leaving the tail pipe.

(5) Further control of exhaust emissions can be achieved by use of the manifold thermal reactor (see figure 4.9). This is an insulated manifold of large volume into which air is injected to provide a high temperature reaction zone for the exhaust gases.

(6) The catalytic converter in figure 4.10 can be used to convert carbon monoxide and hydrocarbons in the presence of air into carbon dioxide and water. Catalysts however are not so effective when there are rapid variations in the gas flow, temperature and gas composition. A further factor that makes the use of catalytic converters difficult is the rapid poisoning of the catalyst by lead compounds.

(7) Oxides of nitrogen can be partially controlled by recycling part of the exhaust gases back into the engine (figure 4.9).

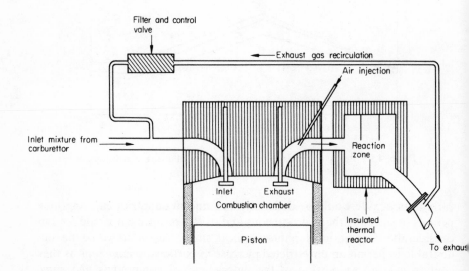

Figure 4.9:   Manifold thermal reactor and exhaust gas recirculation system
(diagrammatic)

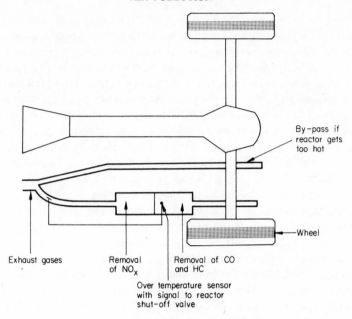

By-pass if
reactor gets
too hot

Wheel

Exhaust gases

Removal
of NO$_x$

Removal of CO
and HC

Over temperature sensor
with signal to reactor
shut-off valve

Figure 4.10: Catalytic reactor system

(8) Filters can be used in the exhaust system for the removal of particulates. Considerable success has been achieved in the removal of inorganic lead particulates using both aluminised stainless steel wool and fibre glass filters.

The inclusion of lead alkyls in petrol has caused a great deal of controversy. They are used as a method of increasing the octane rating of a fuel. This enables petrol engines to be run at high compression ratios giving more power for a fixed amount of fuel. If an engine designed to run on 99 octane fuel was detuned to run on 91 octane, a power loss of 10 per cent would be experienced and there would be an increase in fuel consumption of at least 5 per cent. Among the arguments used in favour of removing lead from petrol is the effect of it upon catalytic converters. These however have their problems in any case where long-term maintenance-free use is required and it may therefore be more feasible to retain some lead alkyls in petrol and institute the use of one of the lead filters described above. Although extensive concern has been shown over the health effects of lead in industrial atmospheres it is doubtful whether it is harmful at the level at which it exists in urban air. Once again this illustrates some of the uncertainties in this field as the effects on man of exposure to very low concentrations of lead over many years is unknown.

If operated correctly the diesel engine poses less of a pollution problem than the petrol engine. In practice however they are often used under excess load and to cope with this, high fuel/air ratios are used. Direct injection of the fuel into the combustion chamber is the feature about the diesel engine that could render it relatively pollution free. A high compression ratio is used as a method of heating the air sufficiently to ignite the fuel and the rate of fuel injection is varied to change the power output. Provided this input is limited to a maximum value, the smoke emissions can be kept within acceptable limits. The advantages of the diesel engine when in correct adjustment, include an emission of only one tenth the carbon monoxide of a petrol engine and slightly less hydrocarbons. There are no blow-by gases as only air is contained within the cylinder on the compression stroke. Furthermore there is little evaporation as the fuel is less volatile and a closed injection system is used for this fuel. As the diesel engine operates at high air/fuel ratios, however, high emissions of oxides of nitrogen occur which are particularly obnoxious where photochemical smog is a possibility.

Emissions from aircraft are again carbon monoxide oxides of nitrogen, hydrocarbons and particulates. Although these are predominant when looking up at an aircraft, by the time dispersion has taken place, levels on the ground are very low. Measurements at ground level at a busy international airport indicate average levels comparable only to a moderately busy street. Frequently the approach traffic and the other airport services give rise to more emissions than the aircraft themselves.

Industrial emissions and their control

The industrial sources of air pollution include electricity generating stations (coal and oil fired), iron and steel works, gas works, lime kilns, ceramic works, sulphuric acid, nitric acid and fertiliser plants, aluminium smelters, copper furnaces and many other smaller industrial processes. In the United Kingdom, for example, there are roughly 30,000 commercial enterprises utilising one or more of these processes and with some form of air pollution problem. These sources generate a range of materials which include both general pollutants and those specific to the processes involved and hence are identifiable. Industry is also the leading producer of particulates, although to be fair a large proportion of smoke emission is from domestic sources. The introduction of domestic smokeless fuels has contributed much to the steady decline in smoke concentrations, although these still give rise to moderate levels of sulphur dioxide in the urban atmosphere. As these are emitted virtually at ground level, however, their relative effect on ground level concentrations can be disproportionately large. The problem of control of air pollution from established processes is thus one mainly of removing particulates: with filters, electrostatic precipitators (where the particles are charged and collected on electrodes) or cyclones (where the centrifugal force exerted on particles in a spinning vortex is employed); or removing noxious gases by the use of scrubbers, adsorbers or afterburners. Careful matching of such

Table 4.1 Control Efficiencies for Industrial Emissions

System type	Removal efficiency (weight per cent)								
	Mineral particulate	Combustible particulate[a]	Carbon monoxide	Hydro-carbons	Nitrogen oxides	Sulphur oxides	Hydrogen chloride	Polynuclear hydrocarbons[b]	Volatile metals[c]
None (flue settling only)	20	2	0	0	0	0	0	10	2
Dry expansion chamber	20	2	0	0	0	0	0	10	0
Wet bottom expansion chamber	33	4	0	0	7	0	10	22	4
Spray chamber	40	5	0	0	25	0.1	40	40	5
Wetted wall chamber	35	7	0	0	25	0.1	40	40	7
Wetted, close-spaced baffles	50	10	0	0	30	0.5	50	85	10
Mechanical cyclone (dry)	70	30	0	0	0	0	0	35	0
Medium energy wet scrubber	90	80	0	0	65	1.5	95	95	80
Electrostatic precipitator	99	90	0	0	0	0	0	60	90
Fabric filter	99.9	99	0	0	0	0	0	67	99

(a) Assumed primarily $\sim 5\mu$
(b) Assumed two-thirds condensed on particulate, one-third as vapour
(c) Assumed primarily a fume $\sim 5\mu$

equipment to the application concerned is necessary. Size of particles for example dictates type of dust arresting equipment employed. Only a few devices will intercept particles smaller than about 2 microns, which are those that remain airborne longest and travel farthest. Nevertheless progress has been steady in developing precipitators and high temperature filter media which make 99 per cent efficiencies attainable for most applications (see table 4.1). Efficiency is directly proportional to cost in all of these applications.

Some processes, however, seem to defy economic control, for example, nitrogen oxides in combustion. Removal of sulphur dioxide from combustion of fossil fuels can only be solved in the long run by removing the sulphur from the fuel and the technology exists to enable this to be accomplished, but cost benefit judgements on the economics of the process can only be made at a national level. Treatment of stack gases to remove sulphur dioxide after combustion is estimated to be 100 times the cost of building tall chimneys (that is, £6 m as compared to £60,000 for a 1 kg/sec sulphur dioxide output rate). Thus for industry a policy of building tall stacks has been adopted which has resulted in a general decline in average ground level concentrations (although total emissions are increasing), due to the small land areas and prevailing wind conditions, which produce acceptable dilution. The Scandinavian countries which now appear in the dispersion pattern for European emissions, however, would not support this policy. The output of sulphur dioxide is probably Britain's major air pollution problem although predictions for 1975 indicate that increasing use of low sulphur fuels is expected to produce a significant reduction in sulphur dioxide emissions.

Air pollution monitoring

Nearly all types of analytical equipment have been used in the monitoring of air pollutants. The choice of sampling technique depends upon the state of the pollutant, that is solid, liquid or gas and methods available include the sampling of gases in polythene bags, the absorption of liquid aerosols onto collection columns and the use of impingers and filters for the collection of solids.

Many simple chemical procedures are available for measuring the common pollutants enabling standardised methods of analysis to be used. Alternatively more sophisticated equipment can be used as is required frequently for an automatic monitoring system. Some of the simplified procedures used are tabulated below.

Sulphur dioxide	Absorption in dilute hydrogen peroxide followed by measurement of total acidity.
Hydrogen sulphide	Measurement of darkness of silver sulphide stain produced on an impregnated filter paper.

Ammonia	Absorption in dilute acid followed by use of Nessler's method.
Fluorides	Collection in water using a perspex impinger. This is then determined colorimetrically as the lanthanum alizarin complex.
Smoke	Use of reflectometer to measure darkness of stain produced on filter paper.
Oxides of nitrogen	Use of the Saltzman method. Absorption in a solution of sulphanilic acid, conversion to the diazo compound which is estimated colorimetrically.
Ozone	Absorption in a buffered solution of potassium iodide and sodium thiosulphate. Back titration of excess with iodine.

Some of the more sophisticated equipment now in use is illustrated by a description of its use in vehicle emission control. The composition of exhaust gases can vary according to the work being carried out by the engine. In order to simulate average emissions, test cycles (figure 4.11) have

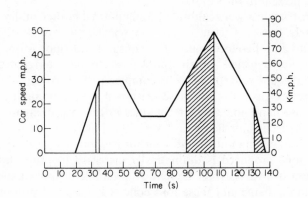

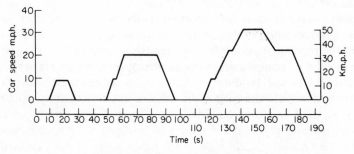

Figure 4.11: European ECE driving cycle

been developed which tend to be specific for different countries or groups of countries. It is of interest that the USA and California standards define limits applicable to all cars whereas European standards are classified according to vehicle size with heavier models permitted to emit a greater total mass of pollutants.

Until 1971 hydrocarbons were generally measured by non-dispersive infra-red (NDIR) analysers. In this instrument infra-red radiation from a glowing filament passes through two identical gas cells and then onto a differential detector. The detector consists of two chambers separated by a flexible diaphragm and filled with the gas to be detected. The radiation transmitted by the tubes is allowed to fall onto opposite sides of the detector. One is the reference cell which does not contain, say carbon dioxide; the other cell contains the air sample. The radiation absorbed by the two halves of the detector produces a heating effect proportional to the intensity. As some of the radiation from the sample tube has been absorbed, the gas in this side of the detector is not heated as much and the difference in pressure causes the flexible diaphragm to distort. This distortion can be monitored electronically as a signal related to the amount of carbon dioxide, in this case, present in the sample.

These instruments were calibrated using normal hexane in dry nitrogen. Unfortunately, however, NDIR is not very sensitive to hydrocarbons other than paraffins and therefore many of the unsaturated smog forming hydrocarbons are not detected. Furthermore background interferences due to carbon dioxide and water need to be compensated for by the use of filters or compensating gas cells. For these reasons it is now becoming standard practice to use flame ionisation techniques in the measurement of hydrocarbons. A hydrogen/air flame is a very hot, blue and relatively ion free mixture. If a trace amount of hydrocarbons is introduced in the fuel the number of ions produced is dramatically increased in proportion to the concentration of hydrocarbon present. If a voltage is applied to a pair of electrodes in the flame and these ions collected, a current proportional to hydrocarbon concentration is obtained. This provides a very sensitive method for determining trace hydrocarbon impurities and is specified as a standard detection for air pollution work as its sensitivity and rapid speed of response make it ideal for detecting 'puffs' of pollution from specific sources. A modification of this method which looks at the luminosity of the flame in the region where sulphur dioxide emits, enables the trace concentration of sulphur compounds, such as mercaptans, which are responsible for much of the 'odour' produced by industrial processes to be tracked down.

Carbon monoxide and carbon dioxide are still measured by NDIR and it can also be used to measure oxides of nitrogen. A new technique that is becoming increasingly used for oxides of nitrogen is chemiluminescence. This is a very sensitive technique which again looks at light emitted only at a specific wavelength which corresponds to electronically excited nitrogen dioxide. This is formed when nitric oxide is reacted with ozone in a small, darkened, low pressure reaction chamber.

$$NO + O_3 \longrightarrow NO_2^* + O_2$$

XV

Nitrogen dioxide can also be measured if it is first pyrolysed to form nitric oxide.

$$NO_2 \xrightarrow{600°} NO + \frac{1}{2} O_2$$

XVI

A similar technique can be used to monitor ethylene as the ozone/ethylene reaction also produces chemiluminescence, but the reverse procedure in which the ozone in the atmosphere is detected by this reaction will probably prove the best instrumental method for monitoring atmospheric ozone levels.

The automatic monitoring of ambient air quality is now becoming an accepted part of public health programmes. There are more problems to this kind of background monitoring than with the monitoring of stack gases for example or vehicle emissions from single sources; the concentrations are generally much lower and a network of analysers rather than an isolated measurement is necessary to obtain data on dispersion and drift patterns. Eventually computer evaluation of simultaneous analytical, meteorological and geographical data will be needed to provide meaningful 'alarms' for potentially dangerous conditions or detect actual high episodes of pollution from single, identifiable sources. In cities prone to photochemical smog, traffic and power generation could also be regulated according to computer predicted limits.

Effects of air pollutants

The toxiological effect of most air pollutants is reasonably well understood but the effects of exposure to mixtures of pollutants, sometimes at low concentrations, over extended periods of time are more uncertain. Epidemiological studies are fequently difficult to carry out and involve a great deal of work if their results are to be meaningful.

A further field of study involves experiments with animals. Studies involving a mixture of pollutants may show an additive effect, amounting to the sum of the effects of each gas or aerosol involved. Alternatively

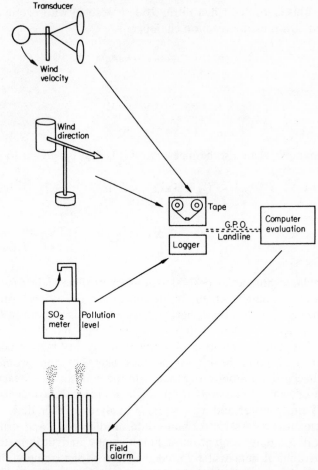

1 Automatic Stations with alarm for high episodes of pollution
2 Dispersion from single sources
3 "Drift" from area sources or industrial
   complexes (directional samples)

Figure 4.12:   Automatic systems

they may produce an effect that is greater than the pure additive effect (synergistic) or less than the additive effect (centagonistic). A further complication is that results may be altered by protective effects and cross tolerances of the animals involved.

Vegetation is more sensitive than animals to many air contaminants and botanical methods have been developed which use plant response to measure and identify contaminants. Sulphur dioxide, hydrogen fluoride ethylene, peroxyacyl nitrates, ozone and aldehydes have all been shown to damage vegetation. However, none of these cases precisely duplicates observed changes in the field, again indicating a lack of knowledge of some of the fundamental chemistry involved. Other variables in this field include changes in the plant growth with age, the effects of light, humidity and temperature, soil composition and variable enzyme systems.

The relationship of air pollution to the ecology of the environment is almost a total mystery. Although it is possible to conceive an ecological cycle in which the specific toxicity of a pollutant for a single species could cause the breakdown of a food chain, the extent to which this occurs is largely unknown.

Some effects of air pollutants such as sulphates, nitrates, etc., could be beneficial to plant growth. It is also worth noting here that among the constituents of photochemical smog are very efficient germicidal agents.

In conclusion, then, air pollution problems must be considered as a complex series of interrelated effects. To understand and attempt to obtain a measure of control over this phenomenon the following points have to be borne in mind:

(1) The natural cycles, the sources and sinks must be identified and their capacities evaluated so that total global tonnages of emissions may be seen in perspective.

(2) The chemistry of air pollution reactions in localised episodes caused by adverse meteorological conditions must be understood so that critical concentrations of possible initiators and reactants may be recognised, estimated and not exceeded.

(3) Maximum permitted levels of exposure should be proposed on toxicological grounds.

Control legislation should thus be adjusted to the lowest of the figures from these three sources. A possible set of limits is given in table 4.2.

Existing legislation and control in the United Kingdom for instance, is based largely on consent although, under the Alkali and Works Regulation

Table 4.2 *Possible Pollution Levels for Typical Species*

Species	Nominal concentration constituting 'pollution' (ppm)	Acceptable background (ppm)
Sulphur dioxide	0.02–1.0	0.2×10^{-3}
Hydrocarbons	$10–50 \times 10^{-2}$	1.0×10^{-3}
Carbon monoxide	1–30	0.15
Carbon dioxide	500	320
Oxides of nitrogen	25 (nitric oxide and nitrogen dioxide)	1×10^{-3}
Ammonia	2.0	$6–20 \times 10^{-3}$
Hydrogen sulphide	$0.5–1.0 \times 10^{-3}$	0.2×10^{-3}
Fluoride	3.0	May be several orders below 'polluted' level

Act of 1906, an Alkali Inspectorate was formed, which sets generally applied standards for sulphur trioxide, smoke, dust, lead and oxides of nitrogen in the region of the emitting source. The Inspectorate, although having wide discretion in the application of these limits can only impose nominal fines for pollution offences. One of the early pieces of legislation on air pollution, the Clean Air Act of 1956, has, on the other hand, been most successful in reducing smoke emissions by enforcing the burning of smokeless fuels, and is undoubtedly responsible for the control of London smogs.

There is very little existing or potential legislation in European countries on control of vehicle emissions in contrast to the extensive Federal legislations in the USA. This reflects an appraisal of the extent of air pollution from these sources on differing geographical and technological scales.

The economic aspects of control or tolerance of pollution are necessarily complex and the final evaluation of the situation is by no means simple. Trends in smoke reduction and sulphur dioxide reduction and the relative absence of major incidents of smog formation or industrial episodes of pollution suggest that the situation in many large industrial cities is steadily improving.

Technologically, then, the methods for evaluating and controlling air pollution at both local and global levels are available and efficiencies of control techniques are improving. The associated economic problem of cost benefit analysis will always present a source of controversy and inevitably represent a political judgement.

Suggested references for further reading

S. S. Burcher and R. J. Charles in *Introduction to Air Chemistry*, Academic Press, 1973

W. Strauss, ed., *Air Pollution Control* (Vol. 1), Wiley-Interscience, 1971

J. O. Leadbetter in *Air Pollution, Part 1 Analysis*, Dekker, New York 1972

W. Bach, *Atmospheric Pollution*, McGraw-Hill, 1972

H. E. Hesketh, *Understanding and Controlling Air Pollution*, Ann Arbor Science Publishers, 1972

CHAPTER 5

WASTE WATERS AND THEIR TREATMENT

G. E. EDEN
Water Research Centre, Stevenage, Herts.

The water-carriage system

In virtually all urbanised societies water is used as a means of carrying away the bodily wastes of the human population. Water is also used in the home for bathing, for the preparation of food, and for the washing of utensils, clothes, and working surfaces. All the waste waters resulting from these uses are discharged to a 'soil drain' and this in turn to a 'sewer'. The sewers serving a community form the 'sewerage system' and the liquid which flows in the sewers is known as 'sewage'. If this sewage is derived only from the household use of water it is known as 'domestic sewage'. In Great Britain the volume of domestic sewage discharged per person per day is about forty gallons.

Water is also used by industry and the resulting waste waters are often discharged to the same sewers as the domestic sewage. In Great Britain the volume of industrial waste water discharged in this way is on average roughly equal to that of the domestic sewage, though the proportions vary considerably from place to place.

In the older cities rain water falling on roads, roofs, and other impermeable surfaces is also admitted to the sewers. This is known as the 'combined' system of sewerage and leads to problems since a heavy rainstorm can increase the flow in the sewer by a large factor in a very short time. In modern urban developments therefore the 'separate' system is used whereby special sewers are provided for dealing with the 'surface waters'. The combined sewer and the 'foul sewer' in separate systems convey the sewage to the sewage-treatment plant or discharge point.

Ingredients of domestic sewage

The composition of a domestic sewage is of course determined primarily by its ingredients. There is firstly the local tap water, which may range in composition from a very 'soft' moorland water (containing perhaps 40 mg/l dissolved solids) to a very 'hard' well water (containing over 500 mg/l dissolved solids).

To this water will be added in the course of use the following pollutants: urine, faeces, paper, soap, constituents of synthetic detergents, scraps of food, earth, grease, cosmetics, and other waste materials. These will be mixed together in the sewer and in the combined system will be diluted in wet weather by surface water containing road grit, salt, oil, soot, and other pollutants washed off impermeable surfaces. The turbulence generated by passage of sewage through the sewerage system tends to break up paper and faeces so that the sewage arriving at the treatment works appears as a grey-brown suspension.

Gross composition of domestic sewage

The particles in sewage range from the colloidal size up to visible coarse matter. Studies using an electron microscope have indicated that 17 to 20 per cent is in a truly colloidal form (table 5.1). On standing (either in the

Table 5.1 *Particle size distribution of suspended matter in sewage*[1]

Size range (diameter)	Description	% by weight (approx.)
> 100 μm	Settleable	50
1–100 μm	Supracolloidal	30–37
1 nm–1 μm	Colloidal	17–20

laboratory or in a sedimentation tank at the sewage works) the settleable solids are removed, leaving the so-called settled sewage, which contains, in addition to the material in true solution, the colloidal and supracolloidal suspended matter. Many of the suspended particles consist largely of living or dead micro-organisms, partly derived from faeces and other pollutants; many micro-organisms can multiply in the sewage, the variety of available organic substances in the sewage serving as substrates for growth. The general properties of a typical strong settled sewage of domestic origin are indicated in table 5.2. The concentration of viable bacteria in sewage is 10^8–10^9 per ml.

This table contains two terms which commonly occur in the literature on pollution, namely COD (chemical oxygen demand) and BOD (biochemical oxygen demand). COD is measured by digesting a sample of the sewage (or other polluting liquid) with strong sulphuric acid, an excess of potassium dichromate, and a catalyst under conditions which ensure that nearly all organic matter is oxidised to carbon dioxide and water; nitrogen compounds remain as ammonia. From the amount of dichromate reduced the amount of oxygen needed for oxidation can be calculated, and this, expressed as mg

Table 5.2 *General properties of a strong settled domestic sewage*[1]

Property	Concentration (mg/l except pH value)
pH value	7.8
Total solids	1,309
Suspended solids	146
COD	670
BOD	370
Organic C	219
Ammonia (as N)	46
Organic N	22

oxygen/litre of the sample, is referred to as the COD. It will be closely correlated with the organic carbon content.

Determination of BOD, in contrast, involves dilution of the sample with a known excess of well aerated river water (or a synthetic dilution water of suitable composition containing a microbial inoculum), followed by incubation of the diluted sample in the dark at 20 °C for five days. Under these conditions as a consequence of the growth of micro-organisms, a proportion of the organic matter is oxidised, while at the same time the dissolved oxygen concentration falls. This reduction in dissolved oxygen, expressed as mg/l of the original sample, is the BOD. Clearly the BOD will always be less than the COD and is usually considerably less.

The BOD test was devised many years ago as a means of assessing the effect of discharging a polluting liquid to a river and, despite many attempts to supersede it, has remained in use for this purpose and also for assessing the concentration of potentially polluting material in a waste water.

Detailed organic analysis of sewage

Attempts to identify in more detail the organic constituents of sewage have met with a fair degree of success. Table 5.3 indicates some of the groups of compounds which have been identified in a whole domestic sewage, while table 5.4 indicates some of the groups which have been identified in a filtered sample of sewage from the same source.

Changes in composition of sewage

The distance from the point at which waste water is discharged to the sewer to a point at which the sewage reaches the treatment works or discharge point may be many miles and the time taken may be many hours. Sewers are designed so that under conditions of average flow they are only partially full and also so that a minimum flow of about 2.5 ft/sec is maintained to prevent deposition of grit. Under these conditions oxygen from the sewer atmosphere is able to dissolve in the sewage at a rate which keeps pace with the rate at which the micro-organisms in the sewage can utilise it by

Table 5.3 *Composition of the suspended organic constituents of a whole domestic sewage*[1]

Constituent	Concentration (mg/l)	Percentage of total organic carbon in suspension
Fats	140	50
Proteins	42	10
Carbohydrates	34	6.4
Anionic detergents	5.9	1.8
Amino sugars	1.7	0.3
Amides	2.7	0.6
Soluble acids	12.5	2.3
Unaccounted for	60	28.6
Organic carbon	211	100

Table 5.4 *Composition of the soluble organic constituents of domestic sewage*[1]

Constituent	mg/l	Percentage of total organic carbon
Sugars	70	31.3
Non-volatile acids	34	15.2
Volatile acids	25	11.3
Amino acids: free	5	3.1
bound	13	7.6
Anionic detergents	17	11.2
Uric acid	1	0.5
Phenols	0.2	0.2
Creatine-creatinine	6	3.9
Unaccounted for	—	15.7
Organic carbon	90	100

respiration, and a low but significant concentration of dissolved oxygen is maintained. The micro-organisms are thus able to proliferate and to begin the process of oxidation of organic matter while the sewage is still passing along the sewer. It has been shown, for example, using radioactively labelled materials, that biodegradable detergents are partially oxidised in this way.

Occasionally it is necessary to pump sewage over a hill or to take it by an inverted siphon across a valley or under an obstruction. Under such conditions the sewage may be retained in a completely filled pipe for a period long enough for the micro-organisms present to utilise all the dissolved oxygen. In the absence of readily reducible salts such as nitrate, reduction of sulphate occurs yielding the objectionable and highly toxic hydrogen sulphide. This substance, liberated as a gas when the sewage next comes into contact with the air, produces dangerous conditions for sewer workers and can lead to a nuisance at and around the sewage works. Also if the hydrogen sulphide is liberated in a normal concrete sewer it may be reoxidised on the walls yielding sulphuric acid which will attack the concrete structure.

Much of the total nitrogen content of sewage originates in urine in the form of urea. During passage through the sewer, and during subsequent treatment at the sewage works, this urea is slowly hydrolysed to ammonia, the reaction being catalysed by the enzyme urease (I):

$$CO(NH_2)_2 + H_2O \xrightarrow{\text{Urease}} (NH_4)_2CO_3$$

I

This hydrolysis must be taken into account by the sewage works chemist in making determinations of the ammonia content of sewage.

Sewage treatment processes

Liquids having the composition shown in tables 5.2 to 5.4 cannot be discharged to a natural water without causing severe pollution unless the dilution available is at least 100-fold. Thus in a country such as Great Britain where large volumes of waste water must be discharged to rivers and estuaries providing relatively little dilution, effective treatment to remove most of the polluting matter is essential.

The processes employed at present are based on the natural processes of sedimentation and biological oxidation which would occur in nature. The processes have, however, been adapted and controlled largely by trial and error over the last seventy years, so that a very high degree of purification can be achieved at a relatively low cost. For example, the present cost of sewage treatment per head in this country is about 0·3p per day. The cost of building a new sewage works at the time of writing is about £20 to £30 per head. These figures may be compared with, say, the current expenditure on newspapers or the capital invested in the family's television set.

Sewage treatment is a series of unit operations and may be usefully studied in the following conventional stages, which are illustrated diagrammatically in figure 5.1.

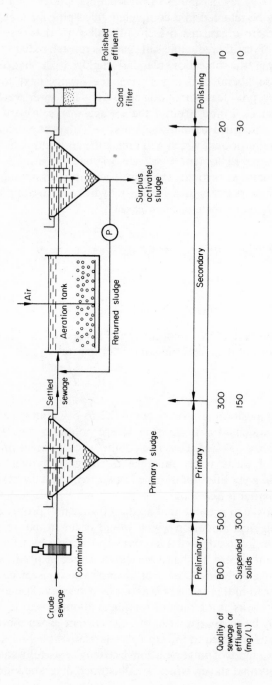

Figure 5.1: Conventional treatment of sewage by the activated-sludge process

1. *Preliminary treatment*

Passage through screens or comminutors (to remove and break up large debris such as rags, paper, and wood), followed by passage through channels in which road grit (but little else) can settle out.

2. *Primary treatment*

Passage through a sedimentation tank having a retention period of several hours in which much of the suspended matter settles out forming the 'primary sludge'. Surplus activated sludge (see below) is often returned to the inlet to the sedimentation tanks.

3. *Secondary treatment*

Biological oxidation and flocculation of most of the remaining organic matter, using either the activated-sludge process or biological filtration.

(a) *The activated-sludge process*   In this process the settled sewage is mixed with a flocculent suspension of micro-organisms (the activated sludge) and aerated in a tank for several hours. During this time most of the organic matter is removed from the sewage by flocculation, adsorption, or oxidation. The sludge, which increases in amount by 5 to 10 per cent during the aeration stage, is removed from the purified effluent in a further sedimentation tank. Most of the sludge is returned to the inlet of the aeration tank, the excess being disposed of by returning it to the inlet to the primary sedimentation tank.

(b) *Biological filtration*   In this process the settled sewage is distributed over the surface of a bed of suitable medium (often slag or rock of about 5 cm diameter), on the surfaces of which bacteria similar to those responsible for purification in the activated-sludge process will develop. Liquid issuing from the base of the filter (which is commonly about 6 ft deep) will contain suspended organic matter (humus), which is removed by sedimentation and disposed of. The effluent is comparable in quality with that produced by the activated-sludge process.

The accumulation of bacteria on the surfaces of the medium is limited by the activities of the so-called scouring organisms—namely fly larvae and worms. These ingest and dislodge the bacterial film, the excreta and dislodged film forming the humus. Since at temperatures below 10 °C the scouring organisms are no longer active, the bacterial film tends to accumulate during the winter and may eventually block the filter before the temperature rises in the spring if the loading of sewage is excessive.

Because of its simplicity in operation biological filtration is preferred for smaller sewage works, and in this country about half the total volume of sewage is treated in this way. The activated-sludge process occupies less space and is more amenable to control and is therefore used in most of the larger works.

(c) *Nitrification*   The process of nitrification (the biological conversion of ammonia through nitrite to nitrate) is often an important feature of both the activated-sludge process and biological filtration, but is of sufficient importance to justify a separate paragraph. Oxidation of ammonia to nitrite is effected by the bacterium *Nitrosomonas* which commonly occurs in activated sludge (II):

$$2NH_4^+ + 3O_2 \xrightarrow{\text{Nitrosomonas}} 2NO_2^- + 4H^+ + 2H_2O \qquad *$$

II

The rate of growth of this bacterium is however rather slow (doubling time about eighteen hours) and if nitrification is to be maintained it is essential that the rate of growth shall be greater than the rate of loss of the organism in the surplus sludge and in sludge particles entrained in the effluent. Expressed algebraically nitrification will occur if

$$t_m = \frac{\Delta S}{S k_M}$$

where t_M is the minimum period of aeration, S is the concentration of sludge solids at the inlet end of the aeration tank, ΔS is the increase in concentration occurring during aeration, and k_M is the growth rate constant of *Nitrosomonas*. The stronger the sewage the greater will be ΔS, and the longer the period of aeration required to ensure nitrification. The effect of temperature is more complex, but the main effect is on k_M, the value of which is approximately doubled for a 10 °C rise in temperature. In practice nitrification is observed to be more difficult to sustain under winter conditions.

The theoretical treatment given above concerns only the conversion of ammonia to nitrite. In practice it is found that if this stage can be successfully achieved the subsequent oxidation of nitrite to nitrate by *Nitrobacter* presents little difficulty (III):

$$2NO_2^- + O_2 \xrightarrow{\text{Nitrobacter}} 2NO_3^- \qquad *$$

III

No doubt rather similar considerations will apply to the process of nitrification during biological filtration, though the complex geometry of a filter does not readily lend itself to theoretical treatment. In general if the

---

* These reactions are simplified, and take no account of the oxygen and nitrogen requirements for bacterial growth.

loading of polluting matter on the filter is moderate and the temperature not below about 10 °C complete nitrification is usually achieved.

4. *Sludge treatment*

Of the organic matter entering the sewage works about half is converted to sludge, the disposal of which represents one of the major problems in sewage disposal as at present practised. In the past, sludge was spread on drained beds of sand or clinker until dry enough to be lifted manually and carted away. This process, which needs a considerable area of land, depends to a large extent on weather conditions and can be offensive. Sludge is increasingly being disposed of by chemical engineering methods, such as filtration in a press or on a rotary-drum filter, followed by incineration; a conditioner such as an aluminium compound or a polyelectrolyte is usually necessary to ensure satisfactory filtration rates. Sludge may also be treated at high temperatures and pressures (220–360 °C and 100 atmospheres) to yield a readily dewatered sludge, but also a highly polluting liquid which must be given further treatment.

Many works employ the anaerobic digestion process, in which sludge is maintained at about 30 °C for about thirty days, during which time about half the organic matter is converted to methane. The resulting thin sludge is odourless and can often be disposed of by direct application to grassland by road tanker. It may also be dried on drained beds without giving rise to nuisance. The methane was at one time regarded as a valuable by-product and is used at many works in dual fuel diesel engines to generate the electrical power needed to operate the works. The economic status of this procedure is of course dependent on the cost of alternative fuels such as electrical power from the national grid.

It is often suggested that sewage sludge should be returned to the land as a fertiliser. In fact in England and Wales about 40 per cent is already so returned, 20 per cent being discharged at sea, and the remainder being dumped on land. It has been calculated that utilisation of all the available sewage sludge would represent only 4.5 per cent of the national requirements for nitrogeneous and phosphatic fertilisers, and less than 1 per cent of the requirements of potassium. The organic matter likewise represents only a few per cent of the national requirement.

Some sewage sludges contain toxic materials discharged to the sewers in industrial wastes and in other cases may contain organisms pathogenic to man or to animals. The use of sewage sludge as a fertiliser must always therefore be viewed with some reservation.

5. *'Polishing' of effluents*

Effluents produced from a sewage of average characteristics by primary and secondary treatment at an efficiently operated works will normally conform to the recommendations of the Royal Commission in regard to quality. (The

Royal Commission on Sewage Disposal suggested in 1912 that an effluent to be discharged to a river providing a dilution of at least 8:1 should have a BOD of less than 20 mg/l and a suspended solids content of less than 30 mg/l.)

This recommendation still forms the basis of many authorisations for the discharge of effluents in the United Kingdom. Unfortunately in many cases the required dilution is not available and it is then necessary for the Water Authority to specify a higher standard of effluent quality. The most common method of improving the quality of an effluent is to pass it through a rapid sand filter; this will remove much of the remaining suspended matter and will yield an effluent having a BOD of 10 mg/l or less and a suspended solids content of about the same value.

Many other methods of achieving about the same degree of purification are employed; these include microstraining (using a very fine mesh rotating screen), storage in maturation ponds, and passage over grassland. The latter methods also provide a very marked reduction in bacterial numbers.

6. *Nutrient removal*

In those parts of the world where accelerated eutrophication (discussed later) presents a problem, for example in Sweden, Switzerland, and parts of the United States, consideration is being given to the removal of phosphates and nitrogen compounds from sewage effluents on the grounds that these substances are limiting in the development of eutrophic conditions. Removal of nitrogen compounds may also be called for in certain other circumstances. It will be recalled that ammonia in settled sewage may or may not be oxidised to nitrate during secondary treatment. Ammonia is objectionable in public water supplies at concentrations above 0.5 mg/l (as N) because it interferes with disinfection by chlorination. Nitrate on the other hand may be unacceptable at concentrations above 10 mg/l (as N) because of its effect in causing methaemoglobinaemia in infants ('blue babies').

(a) *Phosphate removal* This is a relatively simple chemical procedure involving the addition of salts of ferric iron or aluminium, or the addition of lime at a suitable stage in sewage treatment. In Sweden where many new sewage works have been built, all incorporating phosphate removal, the preferred procedure is to use aluminium sulphate in a tertiary treatment stage. A solution of the aluminium salt is added to the effluent from the activated-sludge process, the resulting precipitate flocculated by gentle stirring, and removed in a final sedimentation tank. In this way the phosphate content of the effluent, expressed as P, may be reduced from about 8 to less than 1 mg/l. At the same time the alum floc removes some of the residual suspended and dissolved organic matter, reducing the BOD of the final effluent to as little as 1 mg/l.

If lime is used it may be added to the primary sedimentation stage, the phosphate being precipitated with the sludge as hydroxyapatite $Ca_5OH(PO_4)_3$. This operation raises the pH value to about 9.5. This degree of alkalinity is acceptable since it can be neutralised by the carbon dioxide produced by biological oxidation in the activated-sludge process which follows. The danger of a zone of high alkalinity at the inlet end of the aeration tank is reduced by employing an unbaffled or 'completely mixed' aeration system.

If nitrogen removal also is required, larger amounts of lime may be added to precipitate phosphate and to raise the pH value to at least 10.5 prior to removal of ammonia by air stripping (see below). Lime may also be used to precipitate phosphate from secondary effluent, but here particular care is needed to ensure that the effluent is neutralised before discharge, since the toxicity to fish of any ammonia in the effluent is greatly increased as the pH value increases, the toxic species being un-ionised ammonia (IV):

$$NH_4^+ + OH^- \rightleftharpoons NH_4OH \rightleftharpoons NH_3 \text{ aq.}$$

IV

(b) *Nitrogen removal* Removal of ammonia from sewage, or removal of ammonia or nitrate from sewage effluent are much more difficult procedures than the removal of phosphate and have so far been accomplished on a large scale at only one or two sites. Many schemes have been proposed but only three appear to be feasible; even with these schemes many technical difficulties remain to be overcome, and the processes seem likely to remain expensive.

The process which at the moment seems most likely to prove successful is that involving modification of the activated-sludge process, so that nitrate may be reduced to nitrogen gas. The process is simple in principle but requires careful control. An activated-sludge plant is operated in such a way that all the ammonia is converted to nitrate in the aeration tank. The mixed liquor emerging from this tank then flows into a closed vessel in which it is mixed but not aerated, and into which also a carbonaceous substrate is introduced to maintain the rate of respiration of the sludge organisms. Methyl alcohol is the cheapest source of organic carbon in most areas and is commonly recommended as substrate. Under these conditions the activated-sludge organisms rapidly exhaust the dissolved oxygen and obtain further supplies of oxygen by reduction of nitrate. The sludge is then separated in the usual way and returned to the aeration tank. Many variants of this process have been investigated on the pilot-plant scale. In one variant the denitrification may be carried out as a separate stage after the conventional nitrifying activated-sludge plant using a specially developed denitrifying sludge (figure

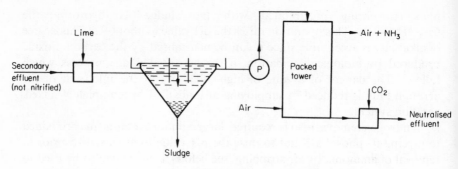

Figure 5.3:   Treatment of sewage by nitrification/denitrification to yield an
effluent low in nitrogen

5.2). The denitrifying organisms may also be developed on a granular support such as activated carbon or gravel. Alternatively the sewage itself may be used as the carbonaceous substrate by creating anaerobic zones in the aeration tank; this process has been demonstrated on the full scale in the UK.

A process which has received a considerable degree of attention in the USA involves removal of ammonia by air stripping. If the sewage or unnitrified sewage effluent is made alkaline using lime to give a pH value of 10.8 or above, the ammonia is mostly converted to the un-ionised form and may be removed by bringing it in contact with a stream of ammonia-free air. Because of the high solubility of ammonia in water very large volumes of air (of the order of 3,000 times the volume of liquid) are required and the contacting tower must be specially designed. The largest installation constructed for this purpose is at the South Tahoe works in the USA. A simplified flowsheet showing removal of both phosphate and ammonia is given in figure 5.3.

Removal of ammonia can be effected by treatment of either the settled sewage or the secondary effluent, provided that the latter is not nitrified. In either case the pH value must be subsequently reduced to a value near neutrality, possibly using carbon dioxide obtained by incineration or calcination of sludges, before further treatment.

In its present state of development this method of removing ammonia is subject to two technical problems. The first of these concerns the deposition of calcium carbonate in the tower; these deposits may appear as soft sludge which can be removed by jets of water, or as a very hard scale which is much more difficult to remove. The reasons for these differences are not fully understood. The second problem is that in cold weather because of the increased solubility of ammonia the efficiency of the process decreases and if the wet bulb temperature of the atmosphere falls below 0 °C, freezing will occur.

Because of the latter problem an alternative process for removing ammonia by ion exchange has been developed. This involves passage of the

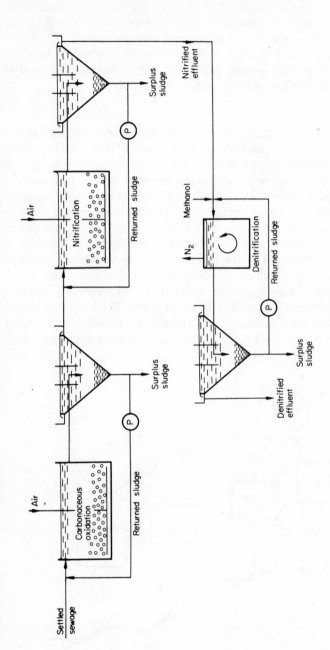

Figure 5.2:   Treatment of sewage by nitrification/denitrification to yield an effluent low in nitrogen

secondary effluent through the mineral clinoptilolite (clino-ptilolite), which is a natural cation exchange material having a strongly selective affinity for ammonium ions. It is regenerated when exhausted by passage of calcium hydroxide solution from which the ammonia may be removed by aeration or by heating. The process is more expensive than aeration but is used at the South Tahoe works as a standby during cold weather.

7. *Water reclamation*

By using the processes described above, waste waters may be purified to a level of purity adequate for discharge to almost any natural water. If this is a river which subsequently serves as a source of water for domestic or industrial use some of the water in the effluent may be said to have been re-used or reclaimed albeit indirectly (figure 5.4). The water supply of London, for example, is largely drawn from the rivers Thames and Lee, and contains on average about 15 per cent of water which has passed through a sewage works. In such a situation there is opportunity for further purification of polluting matter discharged in the effluent, quite apart from the dilution which occurs. Biological self-purification occurs in the river, and during storage in reservoirs, supplementing the further purification by physical, chemical, and biological processes at the water works. There are thus several lines of defence between the sewage works and the next consumer of the water.

Direct re-use, by means of a pipe or the equivalent between the sewage

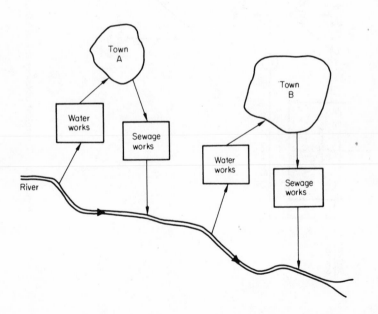

Figure 5.4:   Indirect re-use of water via a river

works and the water works (figure 5.5), potentially presents greater hazards, and has as yet been undertaken on a regular basis at only one site—at Windhoek in South-West Africa. Here effluent from the sewage works is subjected to several further stages of treatment (including storage, chemical treatment and carbon adsorption) and is used to supplement the water obtained from a river source, the proportion of reclaimed water being about one third.

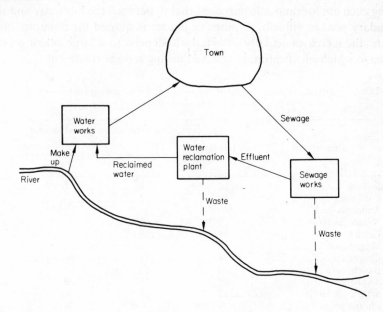

Figure 5.5:   Direct re-use of water

Three problems are of particular significance in connection with the direct re-use of sewage effluent for domestic purposes.

(a) *Viruses*   At water works, chlorination is almost universally employed to ensure absence of pathogenic bacteria. Viruses, however, are much more difficult to remove and to ensure their absence it is necessary to remove as much organic matter as possible (for example by passage through granular activated carbon in columns) before adding sufficient chlorine to yield a residual of free chlorine (hypochlorous acid) after a contact period of one hour.

(b) *Organic matter*   The lack of detailed knowledge of the long-term effects of ingesting organic matter contained in sewage effluent provides a further reason for removing a large proportion of such materials.

E

(c) *Inorganic constituents*   While the suspended and dissolved organic matter in sewage is of paramount importance from the aspect of pollution, it must be remembered that in the course of use inorganic materials also are added to the water, and many of these are not removed or are incompletely removed in the course of sewage treatment. These inorganic pollutants may be of considerable importance in a re-use situation, whether direct or indirect.

Table 5.5 indicates the amounts of inorganic material added to a water during each use for domestic purposes, that is, between the tap water and the secondary sewage effluent. A range of figures is quoted for ammonia and nitrate; the increases actually observed will depend to a large extent on the degree to which nitrification is achieved during sewage treatment.

Table 5.5 *Increase in concentration of inorganic constituents during transition from water through sewage to secondary effluent (all as mg/l)*

Constituent	Increase	Concentration in typical natural water	
		Hard	Soft
Sodium	66	8	3
Potassium	10	2	—
Ammonia (as N)	40–1	0	0.03
Nitrate (as N)	0–30	6	—
Calcium	18	118	3
Magnesium	6	2	9
Chloride	74	17	10
Bicarbonate	100	159	—
Sulphate	28	25	0
Silicate (as SiO_2)	12	18	—
Phosphate (as P)	8	0.09	—
Borate (as B)	1.5	0	0
Total solids	320	375	40

Some of the constituents listed are introduced almost entirely from human sources—for example the nitrogen compounds and the potassium. Others are derived mainly from synthetic detergents; these include silicates and borates. The phosphate is derived in roughly equal proportions from human sources and detergents.

While some of these constituents may be fairly readily removed (for example phosphate) others such as sodium and chloride are not. Whether or not such removal is necessary will depend on the purposes for which the reclaimed water is needed and the degree of dilution. Removal of these inorganic ions would require processes of the kind developed for the desalination of sea waters and brackish waters—for example electrodialysis, reverse osmosis, or even distillation. A point in favour of reclamation of water from effluents rather than from natural saline waters is that the former are available in the areas in which the desalted water is required, so reducing the cost of transporting the product.

Impact of detergents on sewage treatment

Although the subject of synthetic detergents is dealt with in more detail in the next chapter, the increased use of these materials during the last twenty-five years has affected sewage treatment in so many ways that this chapter would be incomplete if it made no mention of this matter. The major effects have been as follows:

1. *Effect on oxygen transfer*
It will be recalled that at the larger sewage works the activated-sludge process is employed and that this involves aeration of the sewage/sludge mixture. In one of the most commonly used forms of this process, air is bubbled through the mixture in deep tanks. The essential characteristic of a detergent is that it possesses surface activity—that is to say it tends to adsorb on to surfaces. The air/water interface around a bubble is one such surface, and as a result of the formation of this adsorbed layer, the transfer of oxygen from air to liquid is appreciably retarded.

This represents in practice a reduction in efficiency of the plant by about 20 per cent compared with its efficiency in the absence of detergents—a considerable economic penalty.

Two further comments should perhaps be added. Firstly, this interference with oxygen transfer (which also shows itself in reduced re-aeration rates in rivers and estuaries) is an effect quite separate from foaming. Secondly, although biodegradable detergents are now generally used, most of the degradation occurs during the biological stage of sewage treatment; thus detergent is still present throughout the aeration tank, and the effect on oxygen transfer, which is not proportional to concentration, is not greatly reduced.

2. *Foaming*
This is the most obvious manifestation of the presence of detergents and has done much to draw the attention of the public to this particular form of pollution, even though foaming is less important in economic terms than the effect on oxygen transfer referred to above. At sewage purification works, foam tends to be produced mainly on the aeration tanks of the activated-sludge plant, and in the days of 'hard' detergents (which were only partially destroyed during sewage treatment) the large banks of foam produced—many feet thick in some places—interfered considerably with the operation of the works by concealing controls and walkways and by covering surfaces with a slippery deposit. Measures are available for suppressing foam by spraying with water or by using anti-foaming preparations, but these are much less

needed now that biodegradable detergents are in general use. Indeed, foam at sewage works is no longer a major problem.

3. *Sludge digestion*

As has already been mentioned, sludge is treated at most of the larger works in the United Kingdom by anaerobic digestion, being transformed in the process to a nearly odourless thin suspension which can be sprayed on to grassland without giving offence. The digestion process is microbiological and is susceptible to poisoning by various substances present in industrial waste waters (notably chlorinated hydrocarbon solvents and metals) and by domestic detergents of the alkyl benzene sulphonate type. These detergents are adsorbed onto the sludge particles settling in the primary tanks. About one-fifth of the detergent entering the works is removed in this way, the remainder passing forward to the aerobic treatment stage, where it is degraded. The concentration of detergent in the sludge is proportional to the concentration in the sewage, and at works receiving a purely domestic sewage (undiluted with industrial effluents) the concentration in the sludge is sufficient to interfere considerably with the digestion process. This is a matter of great practical importance, as a failure of the digestion process involves the works in the disposal of undigested, and hence foul smelling sludge, quite apart from the loss of methane gas. This problem, however, may be overcome by the addition to affected digesters of a calculated quantity of a long-chain primary amine (commercial 'stearine amine') which reacts with alkyl benzene sulphonates to form insoluble complexes.

Both 'hard' and 'soft' alkyl benzene sulphonates are toxic to digestion, so that the introduction of 'soft' alkyl benzene sulphonates has not eased this problem. Other surfactants, such as aliphatic sulphates and sulphonates, and the non-ionic detergents are much less toxic to digestion.

4. *Effects of other constituents of detergent formulations*

All the effects listed above are attributable to the surfactant component of detergent formulations. There are, of course, several other constituents, but on the whole these are without effect on sewage-treatment processes. The phosphates, which as sodium triphosphate ('sodium tripolyphosphate') form the most important inorganic constituent, roughly double the phosphate content of the sewage effluent, but this is of little practical significance in the United Kingdom. In countries where phosphate is to be removed from sewage effluent a rather larger (but not proportionately larger) amount of chemical reagent will be required. It may be noted here that the sodium triphosphate is hydrolysed to orthophosphate during the sewage-treatment processes.

The only other major constituent which calls for comment is borate derived from the perborate included in detergent formulations as a high-temperature bleach. The perborate is reduced during use and passage through the sewer to borate which passes unchanged through the sewage works and appears in the final effluent. Borates are slightly toxic to plant life but it is thought that under conditions in the United Kingdom the agricultural and horticultural use of water derived from rivers containing sewage effluent would not have any adverse effect. A major increase in the use of perborates in detergents would, however, be a matter for some concern and the situation is being kept under review.

Enzymes at present added to certain formulations do not present a pollution problem. They are readily degraded in the washing process and in the sewage and it is unlikely that they would reach the sewage works in an appreciable concentration. Even if they did they would be unlikely to have anything but a beneficial effect since similar proteolytic enzymes play an important part in normal sewage-treatment processes.

Industrial effluents

The foregoing considerations apply primarily to domestic sewage and it is necessary now to consider how the position is affected by waste waters from industry. The volume of water used in industry is of similar magnitude to that used for domestic purposes (water used purely for cooling not being considered), and this water after use is discharged either to sewers or to natural waters.

Since 1st April 1974, responsibility for sewage treatment, pollution control in rivers, and the bulk supply of water for domestic and industrial purposes, has been vested in 10 Regional Water Authorities. Each authority is responsible for setting its own standards of quality for effluents discharged to rivers, whether from its own sewage works or from industrial water users.

1. *Discharge to sewers*
Industry is encouraged to discharge its waste waters to sewers and has a legal right to do so. The Water Authority, however, has the right to set limits to the volume discharged and to the concentration of pollutants in the waste water, and may make a charge which reflects, at least to some extent, the cost of providing sewerage and treatment facilities. The conditions imposed will reflect the authority's own requirements in maintaining an appropriate standard of quality in the effluent discharged from the sewage works. In many cases the industrialist will be required to undertake some treatment of his effluent before discharge to the sewer, particularly where the effluent contains toxic substances.

2. *Discharge to natural waters*
The Water Authority will only accept a discharge to a river or estuary, or into the ground, subject to certain requirements regarding composition and

flow, and these requirements will normally be more exacting than those imposed if the effluent had been accepted into a sewer. While it is not proposed here to go into the question of effluent standards, it will be obvious that an effluent to be discharged to a river serving as a water source will be required to meet a much higher standard than an effluent discharged to an estuary or to the sea.

Classes of industrial effluent

In view of the diversity of industries a rigid classification is impossible, but some of the more important classes and the problems of disposal which they present are dealt with below.

1. Food and drink manufacture
Effluents from such establishments as slaughterhouses, breweries, canneries, beet-sugar factories, and distilleries contain high concentrations of organic matter of natural origin and are usually amenable to treatment by the same methods as domestic sewage. They may thus be treated in special plants installed for the purpose, or in admixture with sewage at the sewage works.

2. Textile processing
Effluents from the textile industries contain substances derived from the washing of the raw materials (wool, cotton, etc.), together with materials used in processing, such as detergents, lubricants, dyeing auxiliaries and dyestuffs. They may be difficult to treat biologically but are commonly discharged to sewers for treatment with domestic sewage.

3. Metal finishing industries
Articles made of metal are often subject to many operations before they are considered saleable or fitted for their purpose. These finishing operations are often conducted in many small and widely-distributed establishments, and involve the production of large volumes of effluents containing rather toxic materials. The potential for pollution is therefore obvious. Some examples of typical processes and the effluents they produce may be helpful.

(a) *Case-hardening* The surfaces of small steel components may be hardened by immersion in molten sodium cyanide. After removal from the molten salt the cyanide is washed off. Cyanide is usually destroyed in such effluents by chlorination in alkaline solution, the cyanogen chloride first formed being hydrolysed to cyanate, a slight excess of chlorine serving as a catalyst.

$$CN^- + Cl_2 \longrightarrow CNCl + Cl^-$$

$$CNCl + 2OH^- \xrightarrow{Cl_2} CNO^- + Cl^- + H_2O.$$

V

Cyanate is harmless environmentally: it cannot be reduced to cyanide. The treatment process may be continuous and controlled automatically using electrodes sensitive to pH value and redox potential.

(b) *Pickling* The removal of surface films from articles made of iron, steel, copper, brass, etc., by immersion in dilute acid is known as pickling and is commonly employed as a preliminary to electroplating, tinning, or galvanising. Effluents from the process may contain acid and dissolved metal and are commonly treated with an alkali such as lime to neutralise the acid and precipitate the metal:

$$H_2SO_4 + Ca(OH)_2 \longrightarrow CaSO_4 + 2H_2O$$

$$Cu^{++} + 2OH^- \longrightarrow Cu(OH)_2$$

VI

The alkali may be added automatically by using a pH controller and the precipitate removed by sedimentation. Waste waters from copper plating from copper sulphate solutions may be treated similarly.

(c) *Electroplating of zinc and cadmium* These elements (and also copper to a small extent) are electrodeposited from cyanide solution. Effluent arising from washing the finished work therefore contains cyanide in the form of the complex salts of the metals. It may however be treated by alkaline chlorination in the same way as effluent containing simple cyanide, a sedimentation stage being added if necessary to remove the metals precipitated as basic salts.

(d) *Nickel plating* Unlike zinc and cadmium, nickel is electroplated from buffered solutions of the sulphate. If treatment is necessary, addition of lime will precipitate basic salts. An important consideration, however, is that if nickel plating is carried out in the same establishment as processes involving cyanide, the two types of effluent must be treated separately. Nickel forms an extremely stable complex with cyanide

$$Ni^{++} + 4CN^- \longrightarrow Ni(CN)_4^{2-}$$

VII

and this complex is not readily treated by chlorination.

(e) *Chromium plating*   This operation involves the use of chromium trioxide solution and yields effluents containing chromate. They may be treated by reduction in acid solution using ferrous sulphate or sulphur dioxide, followed by addition of alkali to precipitate trivalent chromium:

$$Cr_2O_7^{2-} + 3SO_3^{2-} + 8H^+ \longrightarrow 2Cr^{3+} + 3SO_4^{2-} + 4H_2O$$
$$Cr^{3+} + 3OH^- \longrightarrow Cr(OH)_3$$

VIII

One form of anodising (electrolytic formation of a protective film of oxide on the surface of aluminium) also involves the use of chromium trioxide solution; effluents are treated in the same way as those from chromium plating.

4. *Radioactive wastes*

These are subject to special legislation which removes them from the control of local authorities and Water Authorities and places the responsibility for their safe disposal on the Radiochemical Inspectorate of the Department of the Environment. Radioactive wastes arise in the following ways:

  (i) The production of nuclear fuels (extraction of uranium and thorium from their ores, and the subsequent processing and enrichment).

 (ii) The operation of nuclear power stations.

(iii) Reprocessing of nuclear fuels after use in nuclear reactors to recover unchanged fuel, to separate the plutonium manufactured in the process, and to dispose of the accumulated fission products. This reprocessing gives rise to by far the greatest amounts of radioactive waste.

(iv) The production and use of radioactive chemicals for research, in medicine, and in industry.

There are two major principles in the disposal of radioactive wastes:

(a) *Concentration and storage*   Highly radioactive wastes such as those produced in the reprocessing of nuclear fuels are reduced to a relatively small bulk and stored indefinitely. Since the half-lives of many of the constituents are measured in tens of years, storage for long periods is envisaged. The concentrated waste may be stored as a liquid or may be converted to a glass or other inert solid.

(b) *Dilution and dispersal* Less highly active waste waters are treated in such a way that the bulk of the radioactivity is removed. Suitable processes are chemical precipitation (often using a suitable element as 'carrier' for the radioactivity), ion-exchange, or evaporation. The sludge or other residue is stored (or dumped in a form which will not readily disperse), while the liquid, which contains a very low level of radioactivity, may be safely discharged to the environment.

The disposal of radioactive wastes is a complex and highly specialised subject and for further information the recommended text books should be consulted.

Effects of some constituents of industrial effluents on sewage-treatment processes

An account has already been given of the effects of synthetic detergents on sewage-treatment processes. To some extent detergents are a special case, since by far the greatest proportion (about 90 per cent) is used in the home. The housewife is subject to far less stringent controls than the industrialist and cannot, of course, be expected to treat her effluent before discharge.

The general system of administrative control of industrial effluents has already been discussed, and mention has been made of the processes available to the industrialist for removing some of the constituents of his effluents. In this section it is proposed to describe some of the effects industrial effluents would have if discharged to sewers with insufficient treatment. These effects will be taken into account by the Water Authority in setting standards of effluent quality to be met by the industrialist.

1. *Cyanides*
At the pH value of sewage a simple cyanide will be largely hydrolysed to free hydrocyanic acid:

$$CN^- + H_2O \rightleftharpoons HCN + OH^-$$

IX

This substance is a relatively insoluble gas and will tend to come to equilibrium with the atmosphere in the sewer. It has been calculated that, at

F

equilibrium, the maximum safe concentration for men working in the sewer will be exceeded if the concentration in the sewage exceeds 10 mg/l. In many cases this will determine the maximum concentration permitted in the industrial effluent.

At the sewage works cyanides are fairly readily removed by volatilisation as HCN, and by biological oxidation to ammonium carbonate. Cyanides are toxic to fish at concentrations above 0.1 mg/l and the Water Authority will usually wish to ensure that any sewage-works effluent is virtually free from cyanide—a limit of 0.1 mg/l is commonly set and this will be reduced by dilution in the river.

2. *Toxic metals*

Copper, zinc, cadmium, chromium, and nickel are the toxic metals most commonly found in industrial effluents. At the sewage works they will be partially precipitated in the primary and secondary sedimentation tanks and will, therefore, appear in the resulting sludge. If this sludge is to be used in agriculture or horticulture the metal content may be the limiting factor since plants are known to be sensitive to metals such as zinc and nickel in the soil.

In the aerobic processes of sewage treatment metals such as chromium may have an adverse effect, particularly on the nitrification stage. The anaerobic digestion process is subject to inhibition by metals such as zinc and cadmium; here it is possible in an emergency to overcome the toxic effect to some extent by adding sodium sulphide to precipitate the toxic metal.

These metals are toxic to fish and the Water Authority will probably put a limit of the order of 0.5 mg/l on the total concentration in the effluent.

3. *Chlorinated hydrocarbons*

Substances such as chloroform, tetrachloroethane and trichloroethylene are widely used in industry for many purposes, including the manufacture of fine chemicals, the dry-cleaning of clothes, and the degreasing of engineering components. In most of these applications the solvent should not find its way to the sewer, but for various reasons some discharges do occur, and the substances mentioned, and other similar substances, can often be detected in sewage by gas/liquid chromatography.

These substances, being fat-soluble, tend to be concentrated in the primary sludge at the sewage works and can be very inhibitory to the anaerobic sludge digestion process. The most toxic of the group, chloroform, has a measurable effect when present in sewage at a concentration of 0.03 mg/l. The effect of these substances on other treatment processes appears to be negligible and they are removed, presumably by volatilisation, during aerobic secondary treatment.

The examples given are by no means exhaustive, but will it is hoped give some indication of the factors to be considered by a Water Authority in giving consent to the discharge of an industrial effluent to the sewers. Other, more mundane, considerations are no less important. The rate of flow, for example, will tend to reduce the retention time in the sedimentation and secondary treatment stages. The suspended-solids content will determine the additional weight of primary sludge to be treated and disposed of, and the content of biologically degradable organic matter, as measured for example by the BOD, will determine both the additional load to be imposed on the secondary treatment plant and the additional weight of secondary sludge to be dealt with. In general, therefore, the Water Authority will set limits to the concentrations of toxic materials, and will make a charge based on the flow, and on the concentrations of suspended matter and dissolved organic matter.

Effects of pollution on natural waters

The object of treatment of waste waters is normally to render them fit for discharge to natural waters, except in those few instances where water is recovered for re-use. Pollution of water can take many forms, including thermal pollution (discharge of heated effluents), chemical pollution in all its aspects, physical pollution (discharge of suspended matter), and biological pollution (discharge of pathogenic bacteria, viruses and other organisms). The three most important forms of chemical pollution (oxygen deficiency, toxicity, and eutrophication) will be dealt with in more detail.

1. *Oxygen deficiency*
All natural waters contain micro-organisms which require organic matter for growth and oxygen for respiration. The source of this oxygen is that normally dissolved in the water, amounting to 10 mg/l at 15 °C. When bacterial numbers are low, as in a relatively unpolluted stream, the oxygen level is maintained at or near saturation by entry of oxygen from the atmosphere through the water surface. The rate of entry is greatly dependent on the degree of turbulence of the water.

   If biodegradable organic substances are added to the water, as in, for example, a polluting discharge, the numbers of micro-organisms and their respiration rates are increased, resulting in a lowering of the oxygen concentration in the water. This oxygen concentration is paramount in determining the character of a stream. Most fish require a minimum concentration of 3 mg/l (game fish such as trout and salmon considerably more), and even if the oxygen concentration is above the minimum, the toxicity of other poisons may be increased at low concentrations. If the oxygen level falls to zero some bacteria will derive their oxygen requirements by reduction of nitrate, as in the denitrification process already described. If the nitrate becomes exhausted oxygen may as a last resort be obtained by reduction of sulphate, yielding hydrogen sulphide. A river in this condition will be nearly lifeless, black and evil-smelling.

   If, however, the degree of pollution is not too severe the oxygen demand of the organic matter will be met by the resources of the stream (dissolved oxygen supplemented by reaeration) and as the water continues to flow it will be restored to a condition approximating to that above the polluting discharge.

The highly simplified case of a single discharge to a long river without tributaries or abstractions is represented by the so-called 'sag-curve' (figure 5.6). In practice the situation is also complicated by the effects of oxygen

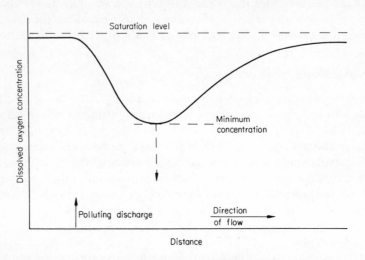

Figure 5.6:   Simplified oxygen sag curve showing minimum concentration
downstream of a single polluting discharge

consumption by mud deposits and by production of oxygen by the photosynthetic activities of aquatic plants, including algae. A Water Authority in considering the authorisation of a polluting discharge will have regard to the effects of this discharge on the oxygen level of the river and will often base predictions on more complex developments of the simple sag-curve. Computer programs for such predictions are available.

2. Toxicity to fish

The appearance of dead fish in a river is, like foaming, one of the more obvious manifestations of pollution. Many factors affect the toxicity of a substance to fish; these include:

(a) Species, age and degree of acclimatisation of the fish.
(b) Temperature, dissolved-oxygen content, hardness and pH value of the water.
(c) Indirect effects of the pollutant on the properties of the water (for example, transparency) or on the organisms serving as food for the fish.

Toxicity is usually expressed as the concentration which (under specified conditions) will cause the death of half a population in a given time. If, for example, the time is 48 hours, the concentration is referred to as the 48-h LC50. In mixtures of poisons the effects are often additive—half an LC50 of one poison plus half an LC50 of another giving one LC50 of the mixture.

The application of these observations to pollution control is not straight-forward since it is not easy to deduce a 'safe' concentration from laboratory measurements of LC50. A considerable margin of safety is usually allowed.

3. *Eutrophication*

This term, which has become more generally familiar in recent years, refers to the enrichment of waters by plant nutrients. In the natural course of events many upland lakes are poor in nutrients and support a sparse but very varied population; such lakes are terms oligotrophic. With the passage of time, erosion and decomposition processes increase the content of nutrients and the lake gradually passes into the eutrophic state characterised by a high biological productivity. This is shown by periodic blooms of algae which render the water unsightly and may cause oxygen deficiency on decay. In general very high concentrations of relatively few species are found. To a lesser extent these considerations apply to rivers also.

This natural progress of eutrophication is accelerated by pollution and much research has been devoted to the elucidation of the controlling factors. In some cases nitrogen compounds and phosphate appear to be limiting and it is for this reason that processes for their removal have been developed. Nevertheless other nutrients may in some circumstances be responsible for promoting blooms of algae.

The sources of nutrients, particularly nitrogen compounds and phosphate, have also been extensively studied. Sewage, even after secondary treatment, contains high concentrations of ammonia (or nitrate) and phosphate unless special steps are taken to remove them. Many rivers, however, receive large amounts of nitrogen in drainage from farmland and though some of this may have originated in artificial fertilisers, even unfertilised land will yield appreciable concentrations as a result of nitrogen fixation in the soil.

Reference

1. H. A. Painter (1971). 'Chemical, physical and biological characteristics of wastes and waste effluents.' *Water and Water Pollution Handbook*, 1, Ch. 7, 329

Recommended further reading

L. Klein (1966). *River Pollution*, vols. 1–3, Butterworth, London

R. L. Culp and G. L. Culp (1972). *Advanced Wastewater Treatment*, Van Nostrand, London

Taken for Granted (1970). Report of the Working Party on Sewage Disposal. HM Stationery Office, London

Notes on Water Pollution. Published quarterly by the Water Research Centre, Stevenage, Herts. (Free on request.)

B. A. Southgate (1969). *Water: Pollution and Conservation*, Thunderbird Enterprises Ltd

J. C. Collins, ed. (1960). *Radioactive Wastes: Their Treatment and Disposal*, E & FN Spon, London

A. L. Downing, ed. (1970). *Water Pollution Control Engineering*, HM Stationery Office, London

CHAPTER 6

DETERGENTS

G. NICKLESS

Department of Inorganic Chemistry, School of Chemistry,
The University, Bristol BS8 1TS

Detergents as sold commercially (either solid or liquid) are complex mixtures of a large number of chemical compounds which include surface active agents, alkaline builders, diluents, water softeners and sequestering agents, stain removers, protective colloids, optical brighteners, acids and solvents. Therefore, any discussion of the pollutant or ecological effects of detergents is inevitably concerned with one or more of the compounds described above and their chemistry in aqueous solution. However, it is of interest to note how 'detergents' arose.

Surfactants

Initially the only effective cleansing agent was soap, which is a salt, usually of sodium, formed by the interaction of fatty acids with alkalis (I):

$$R\,CO_2H + NaOH \longrightarrow R\,CO_2^-Na^+ + H_2O$$

I

R = hydrocarbon chain, normally linear, longer than C_{12} and up to C_{18}.

In commercial practice, natural fats, being the esters of fatty acids and glycerine, are more often employed (II).

$$
\begin{array}{l}
R\!-\!CO_2CH_2 \\
R\!-\!CO_2CH \quad + \quad 3NaOH \longrightarrow 3\,RCO_2^-Na^+ \quad + \\
R\!-\!CO_2CH_2
\end{array}
\qquad
\begin{array}{l}
CH_2OH \\
CHOH \\
CH_2OH
\end{array}
$$

II

Soap is a relatively efficient cleanser but suffers from two major disadvantages:

(1) it is inactivated and forms a scum in hard water, due to the insolubility of the alkaline earth metal salts (especially calcium and magnesium) of fatty acids; and

(2) it decomposes in the presence of dilute acids.

These defects in soap are associated with the presence of the carboxylic acid grouping attached to the hydrocarbon chain. When this was realised, attempts to rectify these faults were started. The term detergent is used rather loosely in everyday life but scientifically describes a synthetic surface active substance which increases the cleansing properties of the medium (usually water) in which it is dissolved. These detergents generally are composed of a 'surfactant' or surface-active agent plus a number of 'builders'. The surfactant lowers the surface tension of the liquid in which it is dissolved by congregating at surfaces and interfaces. The cleansing properties of a detergent arise from its ability to displace the dirt on surfaces by being itself preferentially adsorbed at that site and also helping the dirt to be carried away as an emulsion or suspension (often as a stabilised colloid). The 'builders' help to complex the calcium and magnesium ions in solution so that they do not interact with the surfactant. Also parts of the 'builder' maintain a proper or requisite pH for washing, as well as helping to take the dirt into suspension.

Surfactants are usually a combination of a group which is polar and hence capable of being soluble in water (such as SO_3^-, R_4N^+, $CO.O^-$ or ^-OH) plus a group which is soluble in oil or lipid materials—these are usually long chain alkyl or aryl type groups. Therefore a soap of general formula $R.CO_2^-Na^+$ can be regarded as a simple surfactant. In other words, detergency may be broadly defined as any procedure for the removal of soil and dirt (which in turn may be defined as 'matter in the wrong place') from the surface of a solid by means of a solution.

Therefore, with the introduction of other hydrophilic groups into the fatty acid chain, for example, the sulphonic acid group $—SO_3H$, the sulphur atom being attached directly to a carbon atom in the chain, or the sulphate ester group, $—O—SO_3H$ where the sulphur atom is linked to a carbon atom via an oxygen atom, the first real synthetic surface active agents were manufactured.

The first commercially viable product, Turkey Red Oil was prepared by the low temperature sulphation of castor oil—unfortunately other side reactions took place making it impossible to obtain a product of reproducible quality and specification. Towards the end of the nineteenth century commercially important surface active sulphonic acid materials were synthesised, this route being the basis of the Twitchell process for splitting fats. It involved the reaction of oleic acid and sulphuric acid at high temperatures to produce III.

$$C_{17}H_{33}(SO_3H)COOC_{17}H_{33}COOH$$

III

Such acids were rapidly replaced by more effective products, for example, that formed from the sulphonation of a mixture of naphthalene and oleic acids (IV).

$$HOOCC_{17}H_{33}C_{10}H_6SO_3H$$

IV

No one was quite sure of the substitution positions but the material was a good cleanser, but although the carboxylic acid group was not eliminated, its defects were partially overcome by the sulphonic acid group present.

Reychler in 1913 described his work on aqueous solutions of hexadecane sulphonic acid and its salts. In this and subsequent publications these compounds were shown to be surface active stable to hard water and even mineral acid solution. Unfortunately, no commercial processes were developed for their large scale production since the raw materials and process were too costly.

From these simple beginnings a vast range of surface active agents have been synthesised, studied and used in almost every class of cleansing operation. Indeed by 1953 the total poundage of 'as sold' synthetic detergents exceeded the total soap poundage and since then soap has suffered further serious economic decline. It is interesting to note, however, that the total production of cleansers per capita has remained essentially unchanged.

The major classes of surface active agents which have emerged are classified best as (a) anionic, (b) cationic, (c) non-ionic, and (d) ampholytic types. The International Standardisation Organisation has given the following definitions for each class:

(1) *Anionic* A surface active agent which has one or more functional groups, which ionises in aqueous solution to form a negatively charged organic ion responsible for surface activity.

(2) *Cationic* An agent which has one or more functional groups which ionises in aqueous solution to form positively charged organic ions.

(3) *Non-ionic* A surface active agent which does not produce ions in solution, the solubility in water being due to the presence of functional groups which have a strong affinity for water.

(4) *Ampholytic* An agent having one or more functional groups which, dependent on conditions of medium, may be ionised in an aqueous solution to produce a compound of an anionic or cationic surface active agent.

Typical examples of different types of detergents are given in table 6.1.

In the 1914–18 war, Germany suffered from a shortage of fats and efforts were directed towards the production of synthetic detergents including Igepon A and Igepon T. These were respectively:

    (*a*) the condensation product of chloride of oleic acid and hydroxy-ethane sulphonic acid (sodium salt) (V),

$$C_{17}H_{33} . COOCH_2 . CH_2SO_3^- Na^+$$

V

and (*b*) the condensation product of oleic acid chloride and N-methyl-taurine (sodium salt) (VI).

$$C_{17}H_{33} . CON(CH_3) . CH_2 . CH_2SO_3^- Na^+$$

VI

At the same time, alkylarylsulphonates were produced in the United States and these proved to be the 'breakthrough' for the large scale replacement of soap. In the USA, alkylbenzenesulphonates were used whilst in the United Kingdom alkyltoluenesulphonates were produced; however, they gave poor or limited detergent performance.

Anionic types

The first alkylbenzene sulphonates were complex derivatives obtained through a two-stage process:

    (1) a Friedel-Craft condensation reaction between an aliphatic hydrocarbon and benzene. The aliphatic hydrocarbon being derived from tri- or tetrabutylene which usually contained a high proportion of gemdimethyl groups;

and (2) a sulphonation reaction usually using sulphur trioxide followed by neutralisation to produce the sodium salt.

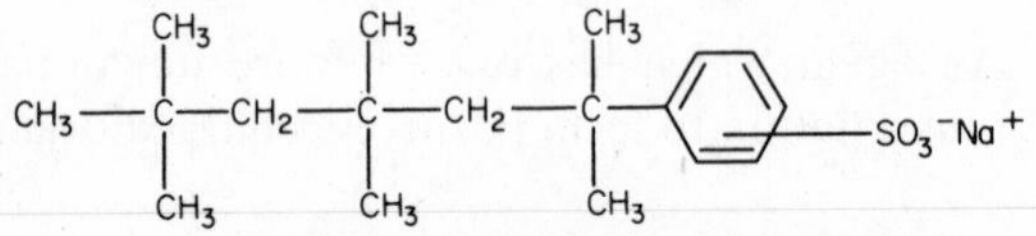

Dodecylbenzene sulphonate (based on tributylene)

VII

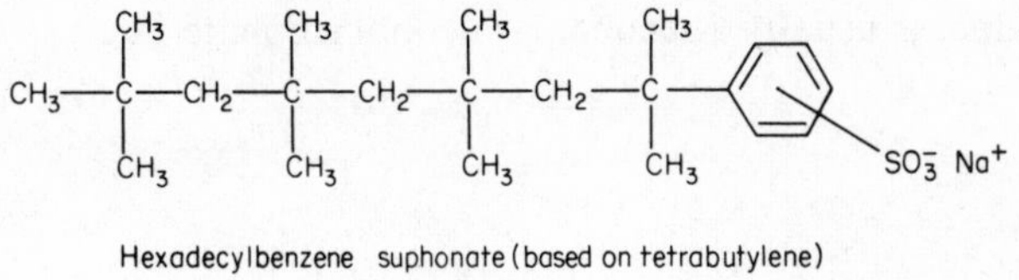

Hexadecylbenzene suphonate (based on tetrabutylene)

VIII

The dodecylbenzene sulphonate (VII) is probably the most widely used of all surface active agents. Most domestic washing powders contain it as a main 'active' ingredient plus a major proportion of domestic washing-up liquids.

Other anionic surfactants include alkyl sulphates and alkylether sulphates. The former are formed by sulphation of a fatty alcohol usually with sulphuric acid. Neutralisation to produce a salt may be with sodium hydroxide, ammonia or an organic base such as triethanolamine. Thus triethanolamine lauryl sulphate is a common basis for hair shampoos whilst the sodium salt is used as a carpet shampoo (IX).

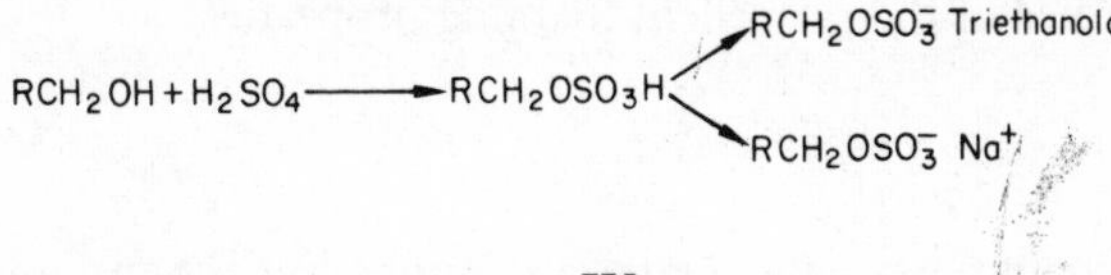

$$RCH_2OH + H_2SO_4 \longrightarrow RCH_2OSO_3H \begin{cases} RCH_2OSO_3^-\ Triethanolamine^+ \\ RCH_2OSO_3^-\ Na^+ \end{cases}$$

IX

Table 6.1 *Types of detergents*

Type	Example
Anionic	Sodium dodecylsulphonate $[C_{12}H_{25}.C_6H_4.SO_3^-Na^+]$
Cationic	Dodecyltrimethylammonium chloride $[C_{12}H_{25}N^+(CH_3)_3Cl^-]$
Non-ionic	Polyethenoxy ethers of alkylphenols $\qquad C_9H_{19}.C_6H_4.O.(CH_2CH_2.O)_nCH_2CH_2OH$ n can be any number usually 10 or more
Ampholytic	Dodecyl-β-alanine (dodecyl aminopropionic acid) $\qquad C_{12}H_{25}NHCH_2CH_2CH_2CO_2H$

Similar products may also be produced by direct sulphation of olefins derived from petroleum feed stocks, usually by the cracking of a paraffinic wax. The final product is usually a secondary alcohol sulphate (X).

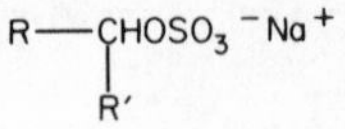

$$R{-}CHOSO_3{}^-Na^+$$
$$\underset{R'}{|}$$

X

Such compounds are important industrial cleaners being prepared on a relatively large scale, for example Teepol is such a compound.

The alkyl ether sulphates are formed by the sulphation of ethoxylated fatty alcohols or alkyl phenols with chlorosulphonic acid.

$$R(OCH_2CH_2)_nOH \; + \; ClSO_3H \longrightarrow R(OCH_2CH_2)_nOSO_3H + HCl$$

XI

$$R(OCH_2CH_2)_nOSO_3H \; + \; NaOH \longrightarrow R(OCH_2CH_2)_nOSO_3^- Na^+ + H_2O$$

XII

The most common of these are based on dodecyl and two to four moles of ethylene oxide ($n = 2\text{--}4$). Such compounds are very mild to the skin and are more water soluble than other alkyl sulphates. Again they are used in shampoos because of their excellent foaming properties.

Cationic types

There are two main types—the quaternary ammonium compounds and long chain amine or amine oxide salts.

Quaternary compounds These compounds are best considered as derived from ammonium salts through replacement of organic groups for the four hydrogen atoms (XIII and XIV).

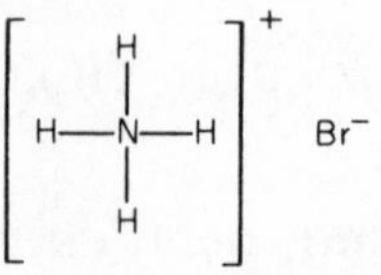

$$\left[\begin{array}{c} H \\ | \\ H{-}N{-}H \\ | \\ H \end{array} \right]^+ \quad Br^-$$

Ammonium bromide

$$\left[\begin{array}{c} C_{10}H_{33} \\ | \\ CH_3{-}N{-}CH_3 \\ | \\ CH_3 \end{array} \right]^+ \quad Br^-$$

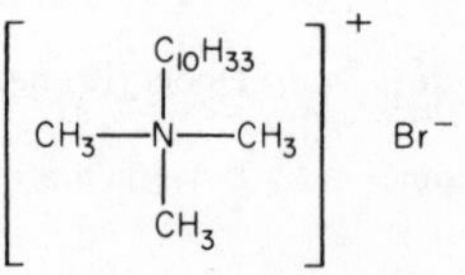

Hexadecyltrimethylammonium bromide

XIII   XIV

Usually such compounds are prepared by the direct addition of for example trimethyl amine and long chain bromide or chloride. Although long chain bromides are easier to obtain and to handle than the chlorides, they are considerably more expensive. Therefore, many studies have been made of the conversion of fatty alcohols to fatty chlorides and high yields can be made without undue operating difficulties.

$$Me_3N + C_{16}H_{33}Cl \longrightarrow (CH_3)_3NC_{16}H_{33}{}^+Cl^-$$

XV

At the present most of the commercially exploited quaternaries have the following generic formula:

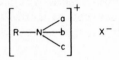

XVI

R is a normal straight chain C_{12}—C_{18}.
a, b and c are methyl, ethyl or benzyl.
X is Cl, Br, methosulphate or ethosulphate.

These products account for a very large proportion of the quaternary cationic surfactants.

Long chain amine or amine oxide salts   Obviously long chain primary amines are protonated in acid solution to form salts for example:

$$RNH_2 + H^+X^- \longrightarrow RNH_3{}^+X^-$$

XVII

Thus R again is usually a normal C_{12} to C_{18} hydrocarbon chain and X is normally acetate or chloride.

Amine oxide salts are prepared by the oxidation of tertiary amines by hydrogen peroxide and may be cationic or non-ionic detergents depending on pH conditions.

$$R_3N + H_2O_2 \longrightarrow R_3N^+O^- + H_2O$$

$$\text{or } R_3NO$$

XVIII

Usually such detergents are used for specific purposes because of expense and instability to heat or alkaline conditions. Thus they are used in textile cleansing whereby producing a soft and 'downy' feel to fabrics. Other uses exploit their germicidal properties so they are employed in dairying and disinfection, food preparation and large scale dish washing.

All these properties are enhanced by the addition of non-ionic detergents but obviously anionic detergents cannot be mixed with cationic detergents because an insoluble fatty complex is produced through double decomposition.

Non-ionic types

The growth in economic terms of non-ionic surfactants within recent years has exceeded all others except alkylaryl sulphonates. This has not been caused by a technical breakthrough but by rapidly decreasing costs of manufacture and a growing appreciation of their practical mertits. Practically any hydrophobic compound which has in its structure a carboxy, hydroxyl, amido or amino group with a free hydrogen attached to the nitrogen atom can be reacted with ethylene oxide to form a non-ionic surfactant; normally an alkaline catalyst is needed.

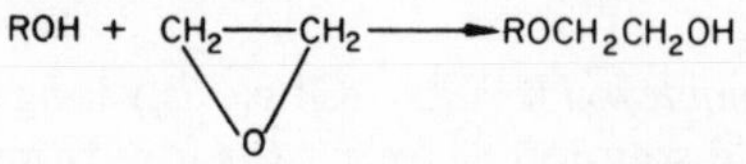

$$ROH + CH_2\underset{\diagdown O \diagup}{\text{---}}CH_2 \longrightarrow ROCH_2CH_2OH$$

XIX

$$ROCH_2CH_2OH + CH_2\underset{\diagdown O \diagup}{\text{---}}CH_2 \longrightarrow ROCH_2CH_2OCH_2CH_2OH$$

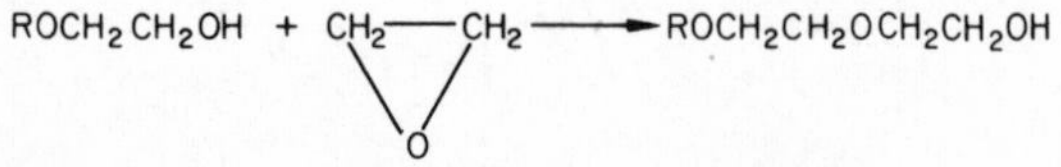

XX

Furthermore, the properties of each compound can be changed considerably by changing the molar proportion of ethylene oxide which is added, in other words by altering the length of polyethenoxy chain. Thus an ethylene oxide nonionic detergent can be obtained at reasonable cost to fit almost any set of requirements. It is easy to build-up chains of fifty or more ethenoxy units, but properties of product depend to a considerable extent on average number of ethenoxy units present. In general, the longer this chain, the higher the solubility in water. The properties of these surface active agents such as wetting power and emulsifying power also change markedly with the number of ethenoxy groups present. But we must be clear that the surfactants in such series are never homogeneous with respect to the length of the ethenoxy chain, and therefore, a fairly wide range of properties are also obtained.

Similar surface active reagents have been prepared by the reaction between alkylated phenols, such as nonyl or dodecyl phenol, in place of the ROH entity. Other materials of interest are 'block type' copolymers where the polymer of some one monomer is joined end to end with the polymer of a second single monomer. They have the generic formula below (XXI):

$$A_n—A—B—B_n$$

XXI

The long polyethylene ether chain forms the hydrophilic portion of a surface active copolymer in XXI where B is $—CH_2CH_2O—$. But the chemical nature of the hydrophobic groups which have been found to have utilitarian purposes would, however, have hardly been predicted. Now polypropylene glycols of molecular weight about 1,000 are not only water-insoluble but can act effectively as hydrophobic chains. Compounds prepared by reacting ethylene oxide with a polypropylene glycol molecular weight 1,200 or higher until a sufficient degree of water solubility has been obtained are extremely effective surfactants.

$$HO(CH_2CH_2O)_n[C(CH_3)HCH_2O]_n(CH_2CH_2O)_nH$$

Polyoxyethylene polyoxypropylene glycols

XXII

There are two major characteristics common to the most non-ionic surfactants which have limited their adaption in many fields. Firstly there is no single nonionic surfactant which can equal the best anionic surfactants with regard to foaming power and foam stability. Certain nonionics have excellent foaming behaviour in the absence of foam—destroying oils and greases, but none can match the anionics over the wide range of actual conditions found in household laundering, dishwashing, etc. Low foam non-ionics have a particular use in machine dishwashing formulations.

The second disadvantage of non-ionics is their liquid nature, since most are viscous liquids or soft pastes and are, therefore, not easily made up to dry, free-flowing powders.

Ampholytic types

These materials which contain both an acidic and a basic function, have been known for many years, but, as a class, they have aroused very little interest. In alkaline media an ampholyte exhibits anionic properties while in acid media, cationic properties. But there is a pH region where the two opposing groups tend to cancel each other out and in such a zone the surface activity is much depressed since the agent has greatly reduced solubility.

The example given in table 6.1 is prepared by adding a fatty amine to an acrylic ester and hydrolysing the resulting amino carboxylic ester (XXIII).

$$RNH_2 + CH_2 = CHCOOCH_3 \longrightarrow RNHCH_2CH_2COOCH_3$$
$$\longrightarrow RNHCH_2CH_2COOH$$

XXIII

Similar compounds have been prepared by using methacrylic esters or crotonic esters in place of acrylic esters. The produce prepared from crotonic esters has the following formula and is available commercially (XXIV).

$$RNH-\underset{\underset{CH_3}{|}}{CH}-CH_2COOH$$

XXIV

As mentioned in the opening remarks surface active agents are only one part of detergents as sold commercially. Table 6.2 depicts the average composition of three detergent formulations. The surfactants which are used in household detergents are chosen on the basis of cost versus performance. If the surfactant is lacking some necessary property, then this property must be supplied by another ingredient of the formulation. The selection of the surfactant depends a great deal upon the type of detergent being prepared, for example liquid types require a highly soluble agent. The polyethenoxy

Table 6.2 *Composition of detergents*
(from P. J. Carr, Ph.D. thesis, Bristol University, 1970)

Component	Detergent 1	2	3	4
	% by weight			
Anionic surfactant	18		5	14
Nonionic surfactant	2	11	2	
Sodium tripolyphosphate	50			45
Sodium carbonate		65	21	
Sodium silicate solids	6	8		
Sodium metasilicate pentahydrate			21	
Sodium chloride			45	⎱ Up to 100 ⎰
Sodium sulphate	14		4	
Sodium carboxymethyl cellulose (c. 65% pure)	< 1	5	1	
Water	10	10		
Brighteners, perfume, foam boosters, etc.	< 1	< 1	< 1	3
Sodium perborate				30

alkyl phenols were among the first products to be used for liquid household formulations. Very high cleaning power but unstable foams are characteristic of these reagents so ammonium or organic amine salts have to be added to stabilise the foams.

Alkyl benzene sulphonates account for the largest quantity of surfactant in the household field, both in Western Europe and United States. They have a low cost and all-round effectiveness but the first highly successful heavy duty synthetic detergents were based on lauryl sulphate.

Additives

In most cleaning operations the surface active agent is used in conjunction with certain non-surface active materials. Such materials referred to as builders or additives, all serve special purposes in the composition; often they form the major portion of the formulation. Additives may be broadly divided into two classes, organic and inorganic, the latter being the class of overwhelmingly greatest economic importance. As described earlier the inorganic builders include (1) alkaline types, (2) water softeners and sequestering agents, (3) stain removers, (4) acids and abrasives.

1. *Alkaline types*
Such agents are mainly sodium, potassium carbonates, sodium silicate and clays. Normally they are added into 'heavy-duty' powder detergents as used in domestic laundry, in quantities up to 60 per cent. They are added simply to ensure that the pH level of the detergent solution does not fall to a level where the surface active agent becomes ineffective. Soda ash and the various forms of alkali metal carbonate are in this category. The silicates take many

different forms, the most important criteria being the ratio of silica (SiO_2) to sodium oxide (Na_2O) in the composition. The most alkaline is ortho-silicate, ratio of $SiO_2 : Na_2O$ of 0.5, formula Na_4SiO_4. But there are meta-Na_2SiO_3 and sesqui silicates, plus a whole range of silicates which are soluble in water and less alkaline than metasilicate, often referred to as 'water glasses'. Their $SiO_2 : Na_2O$ ratio varies from approximately 1.5 to 3.8, they have an effect on hard water which is similar to that of the condensed phosphates in that when added to water they reduce the amount of active agent necessary to produce a stable foam. The effect is usually only observed with an agent of more than fourteen carbon atoms in the hydrocarbon chain.

But all such silicates are good effective corrosion inhibitors, especially in minimising the corrosive action of sulphated or sulphonated active agents on aluminium, brass and other such metals used in washing machines. For this reason they are widely used in household washing and dishwashing formulations.

2. Water softeners and sequestering agents

Within this grouping the polyphosphates are by far the most important compounds, although their dominant position is under attack for reasons discussed later and they face increasing competition from other compounds with similar chemical properties. But as the term suggests condensed phosphate is a descriptive term including within it the pyrophosphates, tripolyphosphates and glassy phosphates (metaphosphates). It was the use of these compounds together with synthetic surface active agents which resulted in the fact that the total poundage of synthetic detergents exceeded the total soap poundage sold in 1953, since they converted 'light-duty' detergents to 'heavy-duty' detergents, which can compete and surpass soap as a cleanser. The major phosphorus-containing component is sodium prepolyphosphate which is used on an enormous scale (about 500,000 tons in the United States for 1968) in both household and commercial laundering situations. Often the tripolyphosphate may constitute 50 per cent of the packet.

These compounds are prepared by the thermal dehydration of ortho-phosphate, the final ratio of Na_2O to P_2O_5 determining the chemical composition of the final material.

$$2Na_2HPO_4 \longrightarrow H_2O \quad + \quad Na_4P_2O_7$$

XXV

$$2Na_2HPO_4 + NaH_2PO_4 \longrightarrow 2H_2O + Na_5P_3O_{10}$$

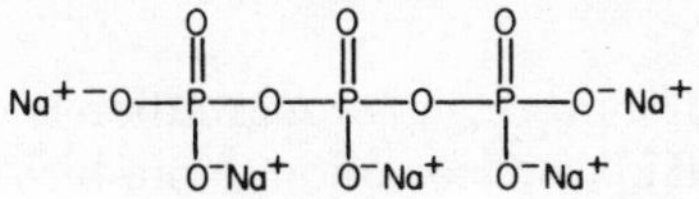

$$Na^+{}^-O-\overset{\overset{\displaystyle O}{\|}}{\underset{\underset{\displaystyle O^-Na^+}{|}}{P}}-O-\overset{\overset{\displaystyle O}{\|}}{\underset{\underset{\displaystyle O^-Na^+}{|}}{P}}-O-\overset{\overset{\displaystyle O}{\|}}{\underset{\underset{\displaystyle O^-Na^+}{|}}{P}}-O^-Na^+$$

XXVI

While the formation of metaphosphate glasses may be written as XXVII.

$$2NaH_2PO_4 \longrightarrow nH_2O + (NaPO_3)_n$$

XXVII

Sodium trimetaphosphate is an example.

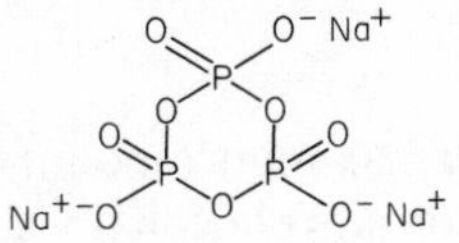

XXVIII

The higher metaphosphates such as sodium hexametaphosphate are probably present in solution as relatively unbranched chains having about 100 or more monomer units per molecule. Active research is being carried out in this area and new structures or suggestions are being proposed frequently. Thus a hexametaphosphate composed of a twelve-membered ring structure of alternate phosphorus and oxygen atoms has been prepared.

However, all the condensed phosphates have three properties in common, which are as follows:

(a) they can sequester metal cations, especially calcium and magnesium ions which are characteristic of hard water;

(b) they can exert peptising and dispersing powers on aqueous suspensions of insoluble solids;

(c) they all revert on hydrolysis to orthophosphate—sometimes called 'reversion'.

Its first two characteristics are closely related to their detergent-builder properties, whilst the measure of their resistance to hydrolysis may give some clue as to how long they can be employed in aqueous solution and hence the length of their effectiveness.

(a) *Sequestering powers* The process of sequestration is usually considered to be the reaction of a metal ion with an anion to produce a soluble complex anion. Sequestration is normally a reversible process and the extent to which it proceeds is described by a thermodynamic equilibrium constant. So with tripolyphosphate the equations can be presented as shown in XXIX.

$$P_3O_{10}^{5-} \; + \; Ca^{++} \; \rightleftharpoons \; CaP_3O_{10}^{3-}$$

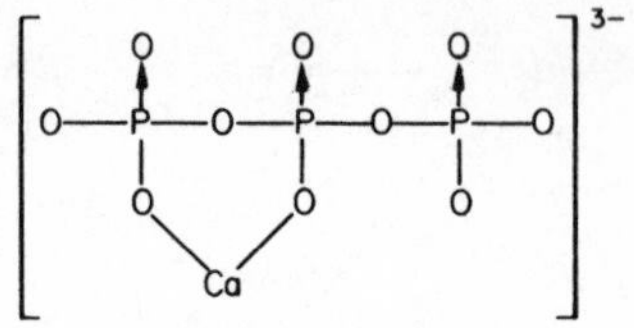

XXIX

What is required is that the maximum amount of calcium ions is sequestered by the minimum amount of polyphosphate, since phosphorus is expensive to produce. Tripolyphosphate is more effective than pyrophosphate, but glassy metaphosphates are generally more effective than tripolyphosphate. Obviously the sequestering power is influenced by pH, temperature, and the presence of other anions in the solution. Also other metals than calcium and magnesium may be sequestered; they include such heavy metals as copper, zinc or iron. Thus the formation of a scum of an insoluble salt with the surface-active agent is prevented, thus stopping any deposit forming in the clothes. Pyrophosphate effectively sequesters magnesium ions, but with calcium ions the initially soluble complex breaks down especially at the higher temperatures of cleansing to insoluble calcium pyrophosphate.

(b) *Dispersing action* The condensed phosphates possess the unusual property of being able to peptise or suspend clays, pigments or other finely divided solids in aqueous solutions. The action is due partly to the sequestering power which locks up calcium or other polyvalent metallic cations, which in turn are strong flocculating agents for negatively charged colloids. The other part of the action is due to their behaviour as surfactants despite their fairly large surface tension in aqueous solutions. So all condensed phosphates can exert a cleansing action on, say, cotton even without a surface active agent present.

(c) *Reversion*   All condensed phosphates hydrolyse in aqueous solutions eventually to orthophosphate, and in all cases hydrogen ions are produced so that the pH of the solution falls.

$$2P_3O_{10}^{5-} + H_2O \longrightarrow 3P_2O_7^{4-} + 2H^+$$

XXX

$$P_2O_7^{4-} + H_2O \longrightarrow 2PO_4^{3+} + 2H^+$$

XXXI

The rates of hydrolysis are dependent upon a large number of factors but certain facts are common to all three classes of condensed phosphate. Thus a rise in temperature increases the rate of reversion, but reversion is slower in alkaline solutions than in neutral or acid conditions. Also metallic ions may be divided into two classes, the first of these include the alkali metals and organic amine cations which tend to increase the stability of phosphates especially those of the glassy form. The second group made up of calcium, magnesium and heavy metal ions decrease the stability of pyro- and tripolyphosphate solutions, especially at the high pH levels.

In general terms, however, the hydrolysis rates are pyrophosphate $<$ tripolyphosphate $<$ glassy phosphate and the rates are separated from each other by an order of magnitude.

Therefore as a compromise, of all the three factors, sodium tripolyphosphate is chosen as the sequestering additive for detergents. However, as mentioned above, the condensed phosphates face increasingly severe competition from organic sequestering agents, of which the two most popular are ethylenediaminetetraacetic acid and nitrilotriacetic acid (XXXII and XXXIII).

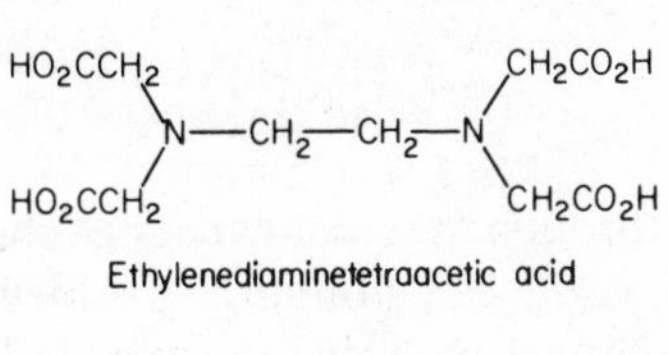

$$HO_2CCH_2 \diagdown \diagup CH_2CO_2H$$
$$N-CH_2-CH_2-N$$
$$HO_2CCH_2 \diagup \diagdown CH_2CO_2H$$

Ethylenediaminetetraacetic acid

XXXII

Nitrilotriacetic acid

XXXIII

The outstanding characteristic of both materials besides their ability to sequester or chelate the heavy metals, calcium and magnesium more effectively than condensed phosphates, is that they do not tend to decompose and hydrolyse in aqueous solution. However, they both tend to bind iron poorly in highly alkaline solution. Thus they can be used as ingredients in liquid shampoos or pastes, but naturally because of their involved syntheses, are more expensive than condensed phosphates. However, as social pressure builds up against polyphosphates, we can expect to see more use being made of such nitrogen containing compounds, although they do not come without their own potential pollution problems as well.

3. *Stain removers*
In many of the powder detergents, compounds similar to sodium perborate $[NaBO_2 \cdot 3H_2O \cdot H_2O_2]$ are used as stain removers. Such material has approximately 10 per cent available oxygen which is released during the hot stages of laundering. Other materials such as powdered calcium hypochlorite or organic chlorine-nitrogen containing compounds such as cyanuric chloride have been used especially in scouring powders. However, chlorine-containing bleaches cannot be used on wool or protein fibres whilst peroxide bleaches do not suffer this disadvantage.

4. *Acids and abrasives*
Abrasives such as pumice or feldspar, can often constitute up to 90 per cent of a typical household scouring powder, but they seldom contain more than 2 to 3 per cent of organic surfactants. Other materials such as alkaline carbonates, phosphates or silicates are also added to the pumice to give complete satisfaction of cleaning surfaces.

Pollution problems

It is not surprising that through the great variety of chemicals which comprise detergents there are a variety of pollution problems which have or may possibly arise from the discharge of the residues of laundering or cleansing processes into sewers or water courses. There are two main effects from these discharges and may be conveniently classified as:

(1) the impact of surface active agents on sewage treatment; and

(2) the impact of dissolved inorganic compounds particularly nutrients leading towards enrichment.

1. *Surface active agent impact on sewage treatment*

This problem has already been very well described in Chapter 5 and little need be added at this stage to the discussion.

The two major effects produced by the organic surfactant were (a) foaming, and (b) reduction in efficiency of oxygen transfer.

(a) Foaming was by far the most spectacular effect of detergents reaching sewage works. The first serious case was recorded at the Mount Penn sewage treatment plant, near the town of Reading in Pennsylvania. The area was primarily residential and was served by a separate sewage system with a mechanical-type activated sludge plant for secondary treatment of domestic sewage. In 1947 a large manufacturer of a household non-ionic detergent covered the entire town of 5,000 people distributing one ounce samples of a detergent. The very next day, sewage plant operators reported large volumes of foam on top of each separator. Two days later, the foam, 2 to 5 ft deep, completely covered most of the treatment plant. Later on hindsight, the plant operators knew that since non-ionic detergents are non-foamers, that the samples given out included anionic detergents. The foam continued each day for thirty days but reached a peak after fourteen days. One of the major difficulties was clogging of basins with grease which was dissolved from the sewer pipes. The control measures included dispersal with both streams of water and spraying surface with anti-foaming agents. But use of such agents for general use was considered too costly. This type of problem was not caused by soaps, since the natural background hardness in normal waters is more than sufficient to precipitate all the soap which enters sewers. Thus, 20 ppm of hardness in water will remove at least 40 ppm of soap. But 5 ppm of synthetic anionic detergent can cause a foaming problem in a plant or a stream.

Gradually this problem or situation built up all over the United States and Western Europe, until in the late fifties some sewers were covered in 30 ft of foam. The situation was finally resolved by the use of legislation which minimised or curtailed the quantity of 'hard' or non-biodegradable products in detergents and hence that which reaches the treatment plants. These hard detergents are normally the highly branched types of alkyl benzene sulphates (for example the chain may be based on tetrabutylene) and they were replaced by linear alkyl benzene sulphonates [LAS] or 'soft' detergents which are more easily degraded by the bacteria in the sewage plant.

$$CH_3—(CH_2)_{15}—\text{benzene ring}—SO_3^-Na^+$$

XXXIV

The mechanism of degradation metabolic route is similar in all these compounds; the first and key step is the desulphonation of the ring, followed by oxidation of the alkyl chain, and finally ring cleavage. In the hard detergents, oxidation of the alkyl chain is halted and hence surface activity remains, but in 'soft' detergents the alkyl chain is oxidised completely away.

(b) The emulsions formed by a detergent may be either stable or unstable. Both emulsions are purposely made so that oil will be removed from the system. Unstable emulsions do not add to the sewage plants problems, but stable emulsions can cause trouble. Certain investigations have shown that 200 ppm of synthetics have prevented completely the settling of precipitative solids at all pH values. The grease and oil emulsions formed in sewers prevent the early removal of oil as scum in the sedimentation tanks and hence increase the load in the biological degradation section. The net result is the prevention of the usual coagulation of dispersed materials. The increased load may be as high as 30 per cent in domestic or up to 100 per cent for industrial products.

When water which contains a small amount of detergent is abstracted from a river for purification to a potable supply, the ability of the detergent to deflocculate or break up soil particles is in direct conflict with the properties needed for optimum treatment. But the actual value in ppm of detergent which causes this lack of sedimentation, coagulation and settling is cause for much debate. Some workers have reported values as low as 5 ppm, whilst others have said 100 ppm cause no problem.

Most detergents inhibit the growth or kill many microorganisms; in the digestion process, both 'hard' and 'soft' alkyl benezene sulphonates are involved. Thus there is a considerable penalty to pay for the need of efficiency and no foam. Cationic detergents tend to destroy the anaerobic bacteria, the limiting dilution for killing being between 1 part in 200,000 and 400,000, whereas the non-ionic detergents are practically inert. Normally, the small quantities of cationic detergents in the sewage system could be neutralised by the large quantities of anionic detergents and hence become innocuous. Therefore effective concentrations of cationics do not arise.

2. Enrichment (Eutrophication)

Enrichment or eutrophication is the natural process which involves an increase in the biological productivity of a body of water as a result of nutrient enrichment from either natural or man-made sources. Concern for the environment stems from where the natural aquatic growth processes are rapidly accelerated or increased by these man-made processes. Nearly all receiving waters naturally change from nutrient-poor (oligotrophic) to nutrient-rich (eutrophic) conditions gradually through time. But water pollution by man can greatly accelerate this ageing process. Lake Erie in the Great Lakes has become a much quoted example. The effects are the same whether eutrophication occurs naturally over geological time or by man's activities. They include excess growth of algae and other plant life, depletion of dissolved oxygen which causes fish to die and development of anaerobic zones where bacterial action produces foul odours and problems in water treatment plants such as clogging of filters, hence undesirable tastes in potable water supplies.

Enrichment supports the photosynthetic process where algae and other plant life synthesise new plant material using energy in sunlight plus a range of nutrients. The nutrients which are used in large quantities are carbon, hydrogen, oxygen, sulphur, potassium, magnesium, nitrogen and phosphorus. The growth process needs traces of other materials, both organic and inorganic; it is also affected by temperature, turbulence in water and poisons or toxins in the water.

Measurements of eutrophication and its effects are highly complex; they rely on a summation of chemical, physical and biological approaches. However, clarification of the variables involved and relative importance to a given body of water requires that such measurements can be made over a long period of time, at least years, but relatively few such studies have been made. Of the known variables, the only one which has been made subject to control by man is the supply of nutrients.

Although all nutrient discharges into the aquatic environment contribute to the reservoir of material available to support aquatic productivity, most control programmes tend to emphasise or concentrate on those nutrients which it is thought can limit productivity.

Green plants consume differing amount of nutrients on which they exist but Liebig's Law of the Minimum states that the essential nutrient that is present in the lowest relative amount limits the amount of growth. Of the major nutrients only carbon, nitrogen and phosphorus have received detailed study and of these the evidence strongly suggests that nitrogen or phosphorus can be important as limiting nutrients. Tables 6.3 and 6.4 depict the major sources of nitrogen and phosphorus, their availability for plant growth and the ease with which they may be controlled. A similar table may be constructed for carbon but most of the sources of carbon are thought to be uncontrollable especially since continuous transfer between living and non-living sections of an ecosystem are known as bio-geochemical cycling. Any nutrient with a gas phase such as carbon is important in this continuous flow, and is considered to exhibit 'complete' cycling. Nutrients which do not have a gas phase, for example phosphorus and magnesium, have cycles which are less perfect. Thus they include a 'sedimentary cycle' involving accumulation in, for example, deep lake or ocean water where they are removed or at the least unavailable for aquatic productivity.

Of the two nutrients, phosphorus is generally thought to be the nutrient most frequently limiting productivity in waters. Therefore, much attention has been paid to phosphorus control in municipal sewage, where it occurs in several forms, totalling about 24 mg/litre as PO_4, broken down into 8 mg/litre

Table 6.3 *Sources and availability of nitrogen*

Nitrogen compounds	Availability	Source	Control
Atmospheric nitrogen (dissolved)	Unlimited supply for N_2—fixing organisms Trace elements must be present (Mo, Co)	The atmosphere 25 mg/l	Remove trace elements or nitrogen fixing algae
Nitrate and nitrite	To all plant life	(a) atmosphere (b) drainge (c) sewage effluent	Uncontrollable Uncontrollable Practically uncontrollable
Ammonia	To all plant life	(a) field drainge (b) animal excretion (c) sewage (d) industrial effluent	Uncontrollable Little remedy available Ammonia stripping in sewage plant Control source
Dissolved organic compounds	To bacteria and other algae	(a) drainage (b) decomposition of organic matter by microorganisms (c) sewage effluents	Uncontrollable Uncontrollable Probably impractical

Table 6.4 *Sources and availability of phosphorus*

Phosphorus compounds	Availability	Source	Means of control
Soluble phosphates	To all plants and animals	(a) field drainge (b) sewage effluent (c) industrial effluents	Practically uncontrollable Precipitation during treatment Remove before discharge
Phosphorus in organic matter	To algae and microorganisms	(a) products of microorganisms (b) bacterial and fungal decomposition of organic matter	Remove at source Remove dead materials
Phosphates as insoluble minerals	To all plants and animals by slow release	(a) large amount under storm conditions (b) rock erosion	Probably uncontrollable Uncontrollable

as suspended solids, 6 mg/litre as soluble polyphosphates and 10 mg/litre as soluble orthophosphate. The principal sources of phosphorus are human excrement and detergents. One can hardly stop or prevent the former, corresponding to about 1–1.2 lb of P/person annually, but the contribution from detergents may be as high as 3.3 lb of P/person. Thus detergents tend to be singled out as unique among consumer goods as a source of nutrients. It has been suggested that 30 to 40 per cent of all phosphorus entering the environment comes via detergents.

In the United Kingdom most sewage is treated to Royal Commission Standards in that less than 30 mg/litre of suspended solids and 20 mg/litre of biological oxygen demand is required as being present in the final discharge. Usually it is expected that this effluent be diluted with at least eight times its volume of clean water, so that the amount of residues remaining will not deoxygenate the receiving water to a point detrimental to aquatic life. Often in low flow conditions this desire may not be fully realised; the soluble fractions amounting to two-thirds of the total P are discharged relatively unaffected by the oxidation procedures in secondary treatment to the receiving water, hence are available to support eutrophication. Typically a secondary treatment effluent can contain up to 2–4 ppm of phosphate, which is approximately fifty times that needed to support maximum growth of certain algae. Such dilution factors are difficult to attain in most urbanised situations. In primary or secondary treatment insoluble forms of phosphorus are removed, but as described in Chapter 5 by Eden, advanced waste water treatment methods are becoming available to remove this soluble portion, the situation being most advanced or developed in the United States and Sweden where this problem is most chronic. (See Chapter 5 for technical details of phosphate and nitrogen removal.)

The next point to emerge has been pressure to remove the phosphorus from detergents in the hope to minimise or curtail the amount of phosphorus in sewage discharges. As seen already, domestic waste contains about 8 mg/litre of phosphorus of which about one-half is from detergents. Obviously this still leaves 3 to 4 mg of P per litre remaining in effluent even if detergent phosphates were eliminated. But as yet there is little evidence that algal productivity would be curtailed if the phosphorus is elevated slightly above the threshold level of 0.01 mg/litre. More importantly small amounts of phosphate may be rapidly recycled between dying and growing algae. Therefore, smallish reductions in phosphate input may have negligible effects on eutrophication in general.

But since the objective of any enrichment control programme is to improve water quality, removal of triphosphate may cause even larger problems because of the supposed needs to our developed society. If we eliminate phosphate, to maintain the performance of present day washing powders (presumably the level the housewife wants), a replacement material must be found to carry those duties which tripolyphosphate does so effectively. However, since in most of our rivers, sewage discharge forms a significant proportion in almost all low-flow conditions, then any suggested alternative which is used on a large scale must be studied in depth, especially its likely concentration and effects on fish, animals and humans. With the present concern for the environment, tests will include practical toxicity tests on the product and its decomposition products. Where the product is organic, its bio-degradability in oxygen or oxygen-deficient conditions will have to be ascertained.

The most publicised alternative is trisodium nitrilotriacetate (NTA) which has already been employed in detergents in the United States and Sweden. Thus NTA has been reported as being fairly rapidly degraded during simulated sewage treatment processes, but more recent Swedish field tests have shown reductions from influent to effluent, under varying conditions as low as 60 per cent but also up to 99 per cent. The actual parameters for a trickling filter system was in the influent 2 to 7 ppm and in the effluent 0.2 to 2.2 ppm; while in the activated sludge process, similar numbers were 5 ppm in the influent and 2 ppm in the effluent. These studies have shown that the level from secondary sewage plants does not exceed a few ppm. The degree of pollution which occurs in receiving waters obviously varies from plant to plant but is probably from 5- to 50-fold or more, giving stream levels of NTA of 0.5 down to 0.04 ppm or less. In a study in Ohio, USA, the average levels at the point of discharges was 0.066 ppm, and 0.025 ppm, about 0.5 miles downstream. Further work was carried out on Long Island well waters during the winter of 1970–71 to ascertain whether NTA from detergents was present in the aquifer (which is slightly contaminated with sewage) from which the inhabitants draw their drinking water. In most cases the level was present only to a maximum of 0.12 ppm. Thus very little NTA can be expected to be present in drinking water, provided that NTA is decomposed in the sewage treatment facility. Two basic problems which deserve further investigation are the factors of temperature and heavy metals. When the lower temperatures of winter are reached then undoubtedly there is a lower reduction of NTA. The concern that chelation of NTA by heavy metals would inhibit NTA degradation in river water has been studied in fair detail and may largely be dismissed since decomposition of copper (II)–nitrilotriacetate takes place to release free copper ions as well as the disappearance of NTA itself.

This picture has been clouded, because late in 1970 the United States Government tests with animals showed that a combination of NTA and two heavy metals which are present in waste water could cause a high incidence of fatal injuries. Thus the statement made was:

At doses used, which were considerably higher than would ordinarily be encountered by the human population, the administration of cadmium and methyl mercury simultaneously with NTA to two species of animals, rats and mice, yielded a significant increase in embryo toxicity and congenital abnormalities in the animals studied over the results with the same dosage with metals alone.

Thus although it is not certain that the use of compounds such as NTA will increase the intake or risk of exposure to trace metals, clearly it is desirable to conduct intensive and detailed long term studies in this area.

Therefore all that can be said at this stage is that although tripoly-phosphate in this country is not under the pressure which it is receiving in say, Sweden, rationalisation by the large industries involved will be necessary and decisions will soon have to be made. Such decisions will have to be taken without all the data being available or all the questions having answers, but this is often the case with pollution where a compromise must be made. When other alternatives to tripolyphosphate appear they will have to be at least as good as these reagents but not cause the problems ascribed to phosphates. However, the limited data now available on non-phosphate containing detergents are not in support of the simple statement that elimination of phosphates will significantly reduce the rate of eutrophication. Certainly the 'ecologically-safe' high alkalinity, high carbonate detergents offer no improvement. Thus to exchange phosphorus for nitrogen may not bring distinct advantages and probably more organic type reagents may be the route to proceed in the future—in other words to ape nature and hence cause less of an environmental impact.

Further reading

J. R. Milton and H. A. Goldsmith (1972). *Systematic Analysis of Surface Active Agents*, Interscience, New York

J. L. Moulliet, B. Collie and W. Black (1961). *Surface Active Agents*, Spen

A. M. Schwarts, J. W. Perry and J. Berch (1966). *Surface Active Agents and Detergents*, Volumes I and II. Interscience, New York

A. M. Andrews, ed. (1972). *A Guide to the Study of Environmental Pollution*, Prentice Hall, Int. Inc.

C. G. Wilber and C. C. Thomas (1969). *The Biological Aspects of Water Pollution*, pp 145–151. Thorough discussion of effects of detergents on fish.

J. Prat and A. Geraud (1964). *The Pollution of Water by Detergents*, Organisation for Economic Co-operation and Development, Paris

D. H. M. Bowen (1970). The Great Phosphorus Controversy, *Environ. Sci. Tech.*, **4**, 725

W. T. Edmondson (1970). Phosphorus, Nitrogen, and Algae in Lake Washington after Diversion of Sewage, *Science*, **169**, 690

F. A. Ferguson (1968). A Nonmyopic Approach to the Problems of Excess Algal Growths, *Environ. Sci. Tech.*, **2**, 188

C. J. Stander and L. R. J. van Vuuren (1969). The Reclamation of Potable Water from Waste Water, *Water Poll. Control Fed.*, **41**, 355

Cleaning our Environment: The Chemical Bases for Action (1969). American Chemical Society, Washington D.C.

A. L. Hammond (1971). Phosphate Replacements—Problems with the Washday Miracle, *Science*, **172**, 361

INDEX